Ebola's Curse

Ebola's Curse

2013—2016 Outbreak in West Africa

Michael B.A. Oldstone
Department of Immunology and Microbiology,
The Scripps Research Institute, La Jolla, CA, United States

Madeleine Rose Oldstone
Graduate of the College of Diplomacy at Seton Hall University,
Carlsbad, CA, United States

ACADEMIC PRESS

An imprint of Elsevier

Academic Press is an imprint of Elsevier
125 London Wall, London EC2Y 5AS, United Kingdom
525 B Street, Suite 1800, San Diego, CA 92101-4495, United States
50 Hampshire Street, 5th Floor, Cambridge, MA 02139, United States
The Boulevard, Langford Lane, Kidlington, Oxford OX5 1GB, United Kingdom

British Library Cataloguing-in-Publication Data
A catalogue record for this book is available from the British Library

Library of Congress Cataloging-in-Publication Data
A catalog record for this book is available from the Library of Congress

ISBN: 978-0-12-813888-5

For Information on all Academic Press publications
visit our website at https://www.elsevier.com/books-and-journals

Working together
to grow libraries in
developing countries

www.elsevier.com • www.bookaid.org

Publisher: Sara Tenney
Acquisition Editor: Linda Versteeg-Buschman
Editorial Project Manager: Timothy J. Bennett
Production Project Manager: Stalin Viswanathan
Cover Designer: Christian Bilbow

Typeset by MPS Limited, Chennai, India

DEDICATION

This book is dedicated to those physicians, nurses, and health care workers who were in Africa or came to Africa to combat the 2013—16 Ebola outbreak. Especially to the staff and workers at the Kenema Government Hospital, Sheik Humarr Khan, Pardis Sabeti, Robert Garry, and the many who lost their lives in the battle to engage and control Ebola, and to Doctors Without Borders, who had been engaging Ebola since its initial onset in Central Africa to the present, along with other maladies.

CONTENTS

Michael B.A. Oldstone is a professor at The Scripps Research Institute, where he directs the laboratory of viral immunobiology. He is a member of the National Academy of Sciences and National Academy of Medicine, the recipient of numerous scientific honors and elections to scientific societies. He was a member of the SAGE executive board of the World Health Organization (WHO) and a consultant to the WHO for the eradication of poliomyelitis and measles.

Madeleine Rose Oldstone is a graduate of the College of Diplomacy at Seton Hall University. Her interest and commitment is to world health problems, policy, and diplomacy.

Ebola's Curse, a timely, needed and well-presented book by Michael Oldstone and Madeleine Rose Oldstone, unlocks the mysteries of the largest outbreak of one of the world's most fearsome viruses. What is Ebola? Why did this happen? Here you will find the answers to these questions, while meeting fascinating people thrust into a situation as dramatic as any that could be imagined in a blockbuster novel or movie.

By reading this book you will come to understand why the world was unprepared for the outbreak of such a deadly pathogen as Ebola virus and why it still is. You will gain intimate knowledge of a pathogen that spread like a tsunami over a region of the world that lacked the resources to fight it and beyond. You will meet a group of people (me among them) that by chance were already there to fight another deadly virus. You will find out how in a matter of weeks this small group of doctors, nurses, and scientists were overwhelmed and why this matters. You will meet people that fought with limited resources at hand and became heroes that put the possibility of saving their patients ahead of their own lives. In the end you will gain insights into steps that must be taken to ensure that such a horrific virus outbreak never happens again anywhere in the world.

This book exists because Michael Oldstone is passionate about viruses. Michael has devoted his career to understanding how viruses undermine and manipulate the immune system thereby causing disease. His investigations of lymphocytic choriomeningitis virus (LCMV), easier known as LCMV, have been beacons shining the light of understanding about fundamental concepts of infectious diseases. LCMV infects the common house mouse around the world; its cousin Lassa virus causes a severe disease known as Lassa fever in humans living in Sierra Leone, Nigeria and other parts of West Africa. Ebola and Lassa fever are so similar that even an experienced doctor cannot tell if a patient has one disease or the other. Unlike Ebola, however, Lassa fever does not cause outbreaks, but is continuously erupting usually in small disease clusters. There is another difference. Lassa fever is somewhat less contagious than Ebola. It does not spread as easily from person to person. Until the outbreak in West Africa the full explosive potential of Ebola had not yet been felt.

In late 2013 I had the privilege of working with Michael to develop a proposal to the National Institutes of Health to apply lessons learned about LCMV infection in mice to Lassa fever in humans. Little was known about how Lassa virus undercuts human immunity. Particularly sparse was knowledge in Michael's specialty known as T cell immunity. Patients from across Sierra Leone come all year round to the Lassa fever ward in Kenema, which is located near the borders of Guinea and Liberia. Our team had been working on Lassa fever in Kenema for over a decade. Our new plan was to use the tools Michael had developed over decades of study of LCMV to probe T cell immunity to Lassa virus in humans. As 2013 ended, we did not know that only a few hours drive away Ebola had just begun to spread to people, and that everything would change.

I will close this brief Introduction by noting a fortuitous circumstance that has produced a boon for readers of this book. As it turns out, the only aspect of his life more precious than science to Michael Oldstone is his family. Teaming this eminent scientist (if there were to be a Wikipedia entry for "eminent scientist" Michael Oldstone's picture could surely appear beside it as one of the best examples) with his nonscientist granddaughter ensured that *Ebola's Curse* is devoid of unexplained jargon and inaccessible technical language. Rather, you have before you a real-life drama about a virus that both terrorized and fascinated the world told in a style that flows seamlessly, teaches without being pedantic, and entertains immensely. What follows is the definitive account of the deadliest outbreak of Ebola.

Robert Garry
Tulane University, New Orleans, LA, United States
October, 2016

ACKNOWLEDGMENTS

The authors thank Drs. Robert Garry, Pardis Sabeti, and Kristian Andersen for their input regarding the Ebola 2013—16 outbreak in Sierra Leone and in Kenema Government Hospital. We are also grateful to the staff of Kenema Government Hospital and Dr. Brian Sullivan for providing insights. We thank Pardis Sabeti, Brian Sullivan, and Kristian Andersen for providing photographs used in this book, and Janet Hightower of The Scripps Research Institute (TSRI) for coupling the photographs into figures and for design of the book's cover. We also acknowledge Gay Wilkins-Blade (TSRI), my faithful secretary of over three decades who typed and retyped our manuscript. We thank Phyllis Minick, our editor in La Jolla, for insights and suggestions. Lastly, we thank Timothy Bennett from Elsevier, the Editorial Project Manager, for guiding our manuscript to a finished product.

ACKNOWLEDGMENTS

The light in Sierra Leone's midafternoon heat, then progressive, inter-mittent darkness encompassed Dr. Sheik Humarr Khan (usually called Humarr or Dr. Khan). Khan was wracked with fever, pain, sweat, continued diarrhea, and increasing difficulty in keeping his eyes open. Khan, the best known physician in Sierra Leone and well recognized internationally for his studies and treatment of viral hemorrhagic dis-eases, headed the Kenema Government Hospital (KGH).[1] This small hospital was flooded with Ebola-infected patients. All beds and even open spaces, including floors, contained sick individuals, many vomit-ing and oozing blood potentially spreading virus until they died. Fewer than half the hospitalized patients recovered. The hospital staff was overwhelmed and exhausted, caring for more than 80 patients in a 14-bed facility. Other patients lay outside on the hospital grounds. Before becoming infected, Khan worked 16−18 h a day. His family warned him to leave this center of Ebola misery. But he told his sister, "If I leave, then who will come and fill my shoes."[2] Already the hospital's head nurse, Mbalu Fonnie, and multiple members of the hospital health staff had died. In areas surrounding the hospital, whispers spread that people go in but do not come out alive. Now, after the testing of Khan's blood confirmed a diagnosis of Ebola infection, he was concerned that his illness would further demoralize KGH's staff and frighten their patients. Khan wrote of his fears, "I'm afraid for my life, because I cherish my life. And if you are afraid then you must take the maximum precautions, stay vigilant and stay on your guard."[1] As evening darkened, Khan was taken from his bed, moved into a vehicle, and driven along dirt roads to the Kailahun district in eastern Sierra Leone, 75 miles away, where Doctors Without Borders had set up an Ebola care center. There, after initial stabilization, Khan's medi-cal condition began a rapid decline. The only therapy available was fluids to replace the 6−10 liters he had already lost by diarrhea, vomit-ing, and sweating. A potential alternative, an anti-Ebola antibody (pro-tein made against the Ebola virus) called ZMapp, was stored at this Doctors Without Borders center but never used for treating humans. In a previous experimental study, ZMapp had been effective in treating

Ebola in monkeys, even when provided 5 days into their illness, but its effect on humans was completely unchartered.[3] Nevertheless, Khan was dying as were many natives, local health care workers as well as medical professionals and others from Western countries.

At this Doctors Without Borders center where Khan now awaited help were others infected with Ebola and also gravely sick. Yet, the center had barely enough ZMapp for three or four persons. Despite his eminence, Dr. Khan was not told that ZMapp was available. The choice of which patients received ZMapp lay primarily in the hands of a team at the Canadian company that made ZMapp and members of Doctors Without Borders at the Kailahun treatment center. But they were not the only decision makers. Also involved were representatives from the World Health Organization, Center for Disease Control and Prevention, and National Institutes of Health. Despite the danger of progressive disease as time passed, the health officials deliberated while considering that neither the antibody's therapeutic effectiveness nor its side effects were known. In the end, they decided not to tell Khan about ZMapp or ask if it could be used on him. Instead, the ZMapp was transported to Guinea where two Ebola-infected victims were treated: a volunteer American physician, Kent Brantly, and a volunteer American health worker, Nancy Writebol, both from Samaritan's Purse charity. Later a priest, Miguel Pajares, from Spain was also given ZMapp. The first two survived but the priest died.[4]

Khan's condition worsened. He reflected on his life of wanting to be a physician and contribute to his country's health, his training in Sierra Leone Medical School, medical residency, a rapid ascent as a physician and expert in therapeutics, his many collaborators from the West, especially Pardis Sabeti from The Broad Institute at Harvard, and Bob Garry and colleagues from the Tulane Medical School.[5] He thought of his work in treating those in need, the good life he had, and whether there was a good death. Then Khan's eyes got heavier; no longer could he open them, and he died. Dr. Khan, who treated over 100 Ebola patients at KGH since the first one entered in March, fell ill in mid-July and tested positive for Ebola at KGH on 22 July, traveled to Kailahun and died 29 July at the Doctors Without Borders treatment center.[6]

This book recalls the origin of Ebola, a new infectious disease when first demonstrated during 1976 in Central Africa, and records the local outbreaks throughout Central Africa from 1976 until 2016. However, the book's emphasis is the 2013–16 West African outbreak where Ebola first appeared as an unknown "mysterious infectious disease" in the village of Meliandou in Guinea and spread from there to Sierra Leone and Liberia. The infection and death rates of the single 2013–16 West African outbreak exceeded the accumulated infections and deaths in total of the previous 25 Ebola outbreaks in Central Africa (Fig. 1.1). The extent of this spread in Sierra Leone represented the majority of Ebola cases in West Africa. Sierra Leone's teams of health care workers, virologists, immunologists, and geneticists in the Kenema region and at KGH were on the front line, and their personal stories form a large part of the body of this book. Dr. S. Humarr Khan, a native of Sierra Leone, was head of its clinical Viral Hemorrhagic Diseases Program, and with Pardis Sabeti of the Broad Institute of MIT and Harvard, and Bob Garry of Tulane Medical School and their colleagues was a leading participant in the fight to diagnose, treat, and understand the Ebola outbreak. Pardis Sabeti, an M.D. and geneticist, and her colleagues analyzed the evolution of the Ebola virus, as it spread from patient to patient and within the individual patient. Robert Garry was the program director and organizer for research at KGH, in conjunction with the Sierra Leone Ministry of Health and Sanitation. The personal experiences of these individuals, their highs and lows, run through this book as do the medical catastrophes and economic hardships resulting from Ebola. How Ebola spread, whether the viruses, people or environments differed, and whether the infection could have been controlled are considered in the chapters that follow. The story would not be complete without pointing to the heroes and their accomplishments on one side of the coin, and on the other side, the errors, mismanagement, and those responsible for prolonging the terror of Ebola. An assessment of errors made locally and internationally is necessary so that a disaster like the one Ebola caused during 2013–16 in West Africa never occurs again. Lastly, this book recognizes and grieves for the over 11,000 victims who died during that epidemic, as well as the 40% of the 800 + health care providers who gave their lives in the fight to contain Ebola including Dr. Humarr Khan.

Michael B.A. Oldstone
The Scripps Research Institute, La Jolla, CA, United States

Madeleine Rose Oldstone
Carlsbad, CA, United States

REFERENCES

1. BBC Staff. Profile: Leading Ebola Doctor Sheik Umar Khan. BBC News; 2014. http://www.bbc.com/news/world-africa-28560507 (accessed 30.07.).

2. Fox News Staff. Sierra Leone 'hero' Doctor's Death Exposes Slow Ebola Response. Fox News Health; August 25, 2014. http://www.foxnews.com/health/2014/08/25/sierra-leone-hero-doctor-death-exposes-slow-ebola-response.html (accessed 23.05.16).

3. Qiu X, et al. Reversion of advanced Ebola virus disease in non-human primates with ZMapp. *Nature* 2014;**514**:47–53.

4. Levs J, Wilson J. Miraculous day as American Ebola Patients Released. CNN Health; August 21, 2014. http://www.cnn.com/2014/08/21/health/ebola-patient-release/index.html (accessed 03.03.16).

5. "Profile: Leading Ebola Doctor Sheik Umar Khan." CNN, July 30, 2014. http://www.bbc.com/news/world-africa-28560507 (accessed 05.03.16).

6. Crowe K. Dying Sierra Leone Dr. Sheik Umar Khan never told Ebola drug was available. CBC News, August 18, 2014. http://www.cbc.ca/news/health/dying-sierra-leone-dr-sheik-umar-khan-never-told-ebola-drug-was-available-1.2738163 (accessed 06.03.16).

CHAPTER 1

Ebola's Origin: A Limited but Devastating Viral Hemorrhagic Disease of Central Africa

The name Ebola comes from a corruption of the French word for the river Legbala (as named in the Ngbandi language). This river is the head stream of the Mongala River, a tributary of the Congo River approximately 166 miles long, in the northern part of the Democratic Republic of the Congo (DRC). This former part of the Belgian Congo was then known as Zaire. In 1976, infection by the so-called Ebola virus was first identified in the town of Yambuku,[1] actually located 60 miles from the Ebola River. Rather than stigmatize Yambuku by naming the virus after the town, and thus hamper its economy or reputation, Dr. Peter Piot, now Director of the London School of Hygiene & Tropical Medicine, called the virus by the river's name. This politically correct technique had been in use for naming, as shown by the outbreak of Hantavirus infection, occurring in the Four-Corners region of the United States where Colorado borders New Mexico, Utah, and Arizona.[2] Originally named Hanta Four-Corners virus, which depicted the geographic site where the virus was found, after disapproval by merchants and residents in the area, the pathogen underwent a change of name to Sin Nombre virus (Spanish for no-name virus) to avoid political and economic outfall. Thus Ebola virus, like Sin Nombre virus, joins the list of politically correct viruses.

In the first outbreak of Ebola virus infections, 318 victims were identified, of whom 279 died, a mortality of 88%.[3] This virus was christened Ebola Zaire, basically the same virus strain as the current one in West Africa.[4] Since 1976 all outbreaks of Ebola virus infection have occurred in Central Africa (Zaire, Sudan, Kenya, Gabon, and Uganda) except for that of one person in the Ivory Coast, West Africa in 1994.[5,6] That person was a scientist who became infected after doing an autopsy on a chimpanzee found in the Taï Forest; presumably, blood from that chimpanzee carried the infectious Ebola viruses.[6] The Taï Forest is a national park in Cote d'Ivorie containing one of the

Ebola's Curse. DOI: http://dx.doi.org/10.1016/B978-0-12-813888-5.00001-9

last areas of primary rainforest in West Africa.[7] The area was designated as a park in 1926 and promoted to national park status in 1972. In 1982, it was declared a World Heritage Site due to the breadth of its flora and fauna; five mammal species in the forest (pygmy hippopotamus, olive colobus monkeys, leopards, chimpanzees, and Jentink's duiker) are on the Red List of threatened species. The Taï Forest is approximately 100 km from the Ivorian coast on the border of Liberia between the Cavalla and Sassandra Rivers. The size of the park is 4540 km^2 and altitudes vary from 80 m to 396 m. Taï Forest is believed to be a natural reservoir of the Ebola virus and is the likely source of the 2013 Ebola outbreak, which then spread through Guinea, Sierra Leone, and Liberia in West Africa during 2014−15.[4,7] None of the multiple occurrences in Central Africa, including the original 1976 outbreak, affected more than 425 individuals or caused more than 280 deaths (Fig. 1.1).[5] Yet, by contrast, the 25th outbreak of Ebola in 2013−15 in Northwestern Africa infected over 28,000 people and killed more than 11,000.

The 1976 eruption of Ebola in Zaire in Central Africa provided lessons for how to control future outbreaks of this disease. Unfortunately, those lessons were not well learned or sufficiently applied to control the recent 2013−15 Ebola outbreak that devastated Guinea, Liberia, and Sierra Leone in Western Africa. Yet the Ebola virus that appeared in the first outbreak (called Ebola Zaire) is basically the same virus found in Western Africa in 2013−15.[4] So what are the lessons and their consequences for those afflicted during 2013−15 and in the future?

The year 1976 marked the first recorded case of an Ebola virus infection and occurred in Central Africa. That index person (first case) was treated at Yambuku Mission Hospital for nose bleed and diarrhea, then fever and lethargy, systemic symptoms that resembled diseases such as malaria, yellow fever, and typhoid fever that are common to the region. This index patient came from a rural area and, though not proven, likely became infected initially after hunting and preparing food contaminated by the blood or saliva of an Ebola-infected monkey or fruit bat. Even when fruit bats carry evidence of Ebola virus, they can be clinically healthy and show no signs or symptoms of infection. Although the infected monkeys usually become ill and die, fruit bats do not.

The Yambuku Mission Hospital had 120 beds and the Ebola virus infection spread rapidly from the single index case to other patients in

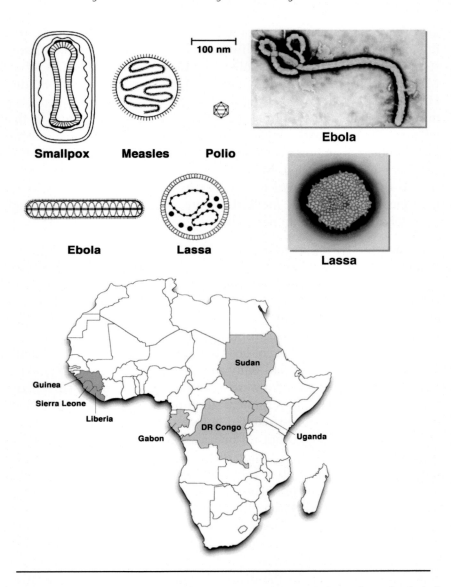

Figure 1.1 Upper: displays comparative sizes of several viruses including Ebola and Lassa discussed in this book. In addition, electron micrograph pictures of Ebola and Lassa are shown. Lower: map of Africa where Ebola broke out in Central Africa (yellow) and recent West African outbreak in blue.

the hospital via use of unsterilized needles, syringes, scissors, and other instruments. At that time and place, hospital instruments were cleaned by simply washing and then rinsing with distilled water before reuse. These practices enabled the viral spread to not only other patients but also health care workers exposed to blood and body fluids from infected patients. Blood, body fluids, saliva, and tears are now known to contain large amounts of infectious Ebola virus. Of the Yambuku Mission Hospital's 17 staff members, 13 became sick, and 11 died. Health care workers and ill patients infected visitors who then transmitted the infection to family members and others on return to their villages. Thus from the single index case, 318 people became infected, and 280 died. The hospital closed when the medical director and three Belgian missionaries died. Many infected individuals and their contacts fled to their home villages out of fear of disease and suspicion of the nonfunctioning Western medical system. Those fleeing often sought traditional therapies from native health healers.[8]

Concurrently, the government of Zaire contacted the United States Centers for Disease Control (CDC) in Atlanta, Georgia, for assistance. Zaire's officials planned to join with and assist a group of international scientists and health care workers to explain and control this outbreak of a lethal hemorrhagic fever whose origin was unknown. The Zaire government of President Mobutu Sese Seko and his Council of Ministers, with the minister of health and the international community, shared information and had daily or at least frequent meetings. More than 70 health care workers were assigned to the field for surveillance and education. The government attempted to quarantine 275,000 people in the area and prohibited commercial plane and boat traffic. Orders were distributed that no one was to leave the villages, and no strangers were allowed to enter them. Four-person teams for health care surveillance, most often led by a physician or nurse, were trained to recognize Ebola hemorrhagic disease. A diagnostic test for Ebola was developed. These knowledgeable surveillance teams visited over 550 villages at least twice over a 2-month period, and a third time for the 55 villages where Ebola had been found. Of great importance were their meetings with the elders and chiefs who headed these villages to impart knowledge about the disease. Team members spoke about the infection's spread and containment methods, the need to avoid contact with sick individuals, and what burial procedure was safe to follow. In many villages, sick patients were placed in outlying huts, following a common practice used earlier for isolation of individuals with small-pox. One family member brought food and water to the isolation hut. When the patient died, the traditional rite of washing and touching the body by family and friends was discouraged. Dead bodies were covered with bleach disinfectant, wrapped in shrouds and buried. Isolation huts housing those who died and their garments were burned.

Hence, the lessons learned were that the signs and symptoms were diarrhea, fever, and lethargy, but such clinical profiles were often confused with other febrile illnesses like typhoid fever, yellow fever, malaria. Infected natives frequently emerged from rural areas then came or were brought to care centers. Transmission occurred by direct contact with infected individuals' blood and body fluids, and spread was amplified by poor hospital practices such as reusing needles, lack of sterilization, and inadequately isolating the sick. Hospital care workers were especially vulnerable and needed protective clothing and adequate health care help. Education and compassion were required to overcome the fear of infection and disease. Training emphasized the importance of isolating infected patients and absolute necessity of disregarding the usual cultural traditions of touching and washing the body of a deceased family member or friend. The involvement of both local and central government as well as international cooperation were obligatory. Formation of health care surveillance groups to monitor local and rural populations was required. Since attempts to physically blockade the migration of people were not effective, as fleeing individuals managed to escape through all barriers erected, the need for education was emphasized repeatedly. Further established were the facts that infection from needle sticks required an incubation period of 6 days for symptoms to appear, and person-to-person contact took a mean of 8 days with a range of 1–21 days for the onset of disease. Therefore a mandatory quarantine period of 42 days (double 21 days) was selected to prevent further contamination.[9]

In 1978, 2 years after the index case appeared, Ebola resurfaced at Tandala Mission Hospital, located about 155 miles from Yambuku. In this instance the disease did not spread. The physician in charge, who had participated in the Yambuku outbreak, suspected that the agent of infection was Ebola virus and isolated the 9-year-old girl who came from a rural village to enter the hospital. Only one other person became infected with Ebola virus, the girl's younger sister.[10]

However, an extraordinary challenge in the control of Ebola was to occur 7 years later in 1995. The 1995 outbreak provided the stringent test of whether the lessons of containment learned earlier in rural areas could also be applied to stop a devastating spread of the infection into the largest city population center in the country recently renamed from Zaire to the DRC. The Kikwit General Hospital in DRC was separated from the capital city by 217 miles and 5 h driving time along well-traveled roads. In Kinshasa the capitol, lived several million susceptible individuals, many in close quarters and in slums.

After lying quiescent in DRC (Zaire) for 18 years and for 16 years elsewhere in central Africa (Sudan in 1979, 34 individuals infected with a mortality rate of 65%), Ebola reappeared in DRC in May 1995. Increasing numbers of patients sick with hemorrhagic fever entered the Kikwit General Hospital.[5,6,11] In short order, many of the patients hospitalized for treatment, their families who accompanied them, and nurses and doctors who treated those sick individuals died. Ebola had been suspected by local physicians who observed similar cases 19 years earlier. As recorded "...When a 36-year-old lab technician known as Kinfumu checked into the general hospital in Kikwit, Zaire, last month, complaining of diarrhea and a fever, anyone could have mistaken his illness for the dysentery that was plaguing the city. Nurses, doctors, and nuns did what they could to help the young man. They soon saw that this disease wasn't just dysentery. Blood began oozing from orifices in his body. Within four days he was dead. By then the illness had all but liquefied his internal organs."[11]

That was just the beginning. The day Kinfumu died, a nurse and a nun who cared for him fell ill. The nun was evacuated to another town, 70 miles to the West where she died—but not until the contagion had spread to at least three of her fellow nuns. Two subsequently died. In Kikwit the disease raged through the ranks of the hospital's staff. Inhabitants of the city began fleeing to neighboring villages. Some of the fugitives carried the deadly virus with them. Terrified health officials in Kikwit sent an urgent message to the World Health Organization. The Geneva-based group summoned expert help from around the globe: a team of experienced virus hunters composed of tropical-medicine specialists, virologists, and other researchers. They grabbed their lab equipment and their protective suits and clambered aboard transport planes headed for Kikwit.

Except for a handful of patients too sick to run away, the hospital was almost abandoned when the experts arrived. Supported by many others who were already there or soon to arrive, the response to the outbreak was headed by virologist Jean-Jacques Muyembe and epidemiologist David Heymann. Both went to work rapidly applying forceful methods to hamper and then stop the spread of Ebola infection.[10,11] They first identified patients who had traveled to the capital, Kinshasa, and isolated them. They began educating the community about the risk of infection and the need for proper containment. Finally came tactics for persuading the community to forego dangerous funeral rites: they must not congregate around or touch deceased family members and friends, not wash out a corpse's mouth, not wash the body, and not clip fingernails or hair. The DRC government tried to cordon off the city to prevent inhabitants from spreading the contagion across the countryside and into the sprawling slums of Kinshasa. The quarantine was only partly effective, as it had been years since there was a functioning government in DRC. The international doctors sent people with bullhorns through the streets pleading with residents to stay at home.

The next strategies were to rapidly identify and isolate those infected with Ebola virus and require health care workers to use protective clothing. Concurrently, health surveillance teams were established to identify and follow locals who were in contact with Ebola-infected patients and to monitor their temperatures twice a day for 3 weeks. Those with fever were isolated until disease was confirmed or excluded. Those testing positive for Ebola were hospitalized. It was essential to educate individuals in the population area at risk on how to protect themselves and their families. Work by several international agencies, namely the Red Cross and the Red Crescent Societies, met with village elders and chiefs to distribute information tailored to local customs, that is, identify tribal remedies and funeral practices to avoid and prepare village leaders to advise members of their villages accordingly. Lastly, protective clothing and gear were needed for those transporting patients to hospitals for medical care and for those performing burial services.

The identification of Ebola virus infection encountered endless difficulties. Specimens were collected and forwarded via the Belgian Embassy to the Institute of Tropical Medicine in Antwerp for evaluation. But the specimens could not be tested there for diagnosis of Ebola, because that institute no longer had the appropriate containment laboratory (BSL-4) for such studies. In Belgium, as elsewhere including the United States, short-term political considerations had reduced funding for surveillance as well as research into infectious diseases. The samples then traveled from Antwerp to the Centers for Disease Control in Atlanta, Georgia, where tests were performed and identified those patients infected with Ebola virus. Locally around Kikwit, the road blockades enforced by DRC military were not effective at confining residents. Regardless of fever surveillance, fear drove both the sick and well villagers to avoid main roads, use forest paths to the Kikwit river, then travel further by boat.[12]

Nevertheless, the restrictions that were enforced prevented the spread of Ebola infections to the large city of Kinshasa. The 1995 outbreak in Kikwit was limited to a recorded 316 Ebola-infected persons with 243 deaths, a mortality of 77%. This number is likely a low estimate of the disease spread and does not include those not identified who were ill and died in rural areas.

As we will see, the subsequent failure to contain the 2013–15 Ebola outbreak in West Africa stemmed from a breakdown of several protocols that should have been followed. For example, in 2014, a quarantine failed at an urban slum in the large city of Monrovia, Liberia.[13] That quarantine had no chance of success, because it was lifted just days after its declaration when armed members of the population clashed with the government forces. As a result, the infected residents and others mixed at will then moved freely in and out of the quarantined area. They avoided checkpoints and used bribery. Further, this population was more mobile with better transportation (cars and motorcycles) and better roads than those dwelling in Central Africa.

By this time the Ebola virus had been characterized as belonging to the family of filoviruses, so-named for their long and thread-like shape that resembled filum, Latin for thread. Fig. 1.1 displays an electron microscopic photomicrograph of the virus emphasizing its length, up to 14,000 nm, varying 800–1200 nm, and its narrow diameter of only 80 nm. The seven genes comprising the viral genome total 19Kb, which is large for a negative single-strand RNA virus, but small for most DNA viruses.

The outbreaks of Ebola since the first report in 1976 to the most recent outbreak in 2014 are shown in Table 1.1 along with the three single cases of laboratory accidents in 1994, 1996, and 2004.

The molecular biology of Ebola virus, the immune response in the human host to the virus, and its infectious nature are not completely

Table 1.1 Displays Chronologically all Ebola Outbreaks Known as to Year, Location, Number of Cases, and Percent Mortality, and Ebola Strain Implicated

Ebola Outbreaks			
Year	Country	#Cases/%Mortality	Virus
1976	Zaire	318/88%	Zaire
1976	Sudan	284/53%	Sudan
1976	England	1/0%	Sudan
1977	Zaire	1/100%	Zaire
1979	Sudan	34/65%	Sudan
1994	Gabon	52/60%	Zaire
1994	Ivory Coast	1/0%	Tai Forest
1995	DR Congo	315/81%	Zaire
1996	Gabon	37/57%	Zaire
1996−97	Gabon	60/74%	Zaire
1996	South Africa	2/50%	Zaire
1996	Russia	1/100%	Zaire
2000−01	Uganda	425/53%	Sudan
2001/2002	Gabon	65/82%	Zaire
2001−02	DR Congo	57/75%	Zaire
2002−03	DR Congo	143/89%	Zaire
2003	DR Congo	35/83%	Zaire
2004	Sudan	17/41%	Sudan
2004	Russia	1/100%	Zaire
2007	DR Congo	264/71%	Zaire
2007−08	Uganda	149/25%	Bundibugyo
2008−09	DR Congo	32/47%	Zaire
2011	Uganda	1/100%	Sudan
2012	Uganda	11/36%	Sudan
2012−13	Uganda	6/50%	Sudan
2013−15	West Africa	28,644/40%	Zaire
2014	DR Congo	66/74%	Zaire

understood. Research is hampered, because Ebola is such a dangerous and severe human pathogen that it cannot be handled in a conventional laboratory but only in the highest-level containment facility known, a so-called Biosafety 4 laboratory (BSL-4). Scientific handlers must be enclosed in specially designed space-like suits under positive pressure. Additionally, the virus intermittently disappears as an infectious agent only to reappear eventually (Fig. 1.1).

Viruses contain either RNA or DNA and are, therefore, categorized as RNA or DNA viruses. Ebola and Lassa, to be described later, are the main viruses that produce hemorrhagic diseases in Africa, and both are RNA viruses. RNA viruses are the only organisms known to use RNA as their genetic material. They replicate their RNA genome in one of two unique ways. First, by either RNA-dependent RNA synthesis (Ebola, Lassa, and most RNA viruses (i.e., measles, influenza, poliomyelitis, etc.)) or, second, by RNA-dependent DNA synthesis, so-called reverse transcription, followed by DNA replication and transcription (retroviruses like HIV).

Importantly, RNA replication is error prone, since this class of viruses does not have a strong proof-reading mechanism that corrects errors by removing wayward or mutated nucleic acids. The enzyme (polymerase) that catalyzes RNA replication has a minimal proof-reading activity. As a result, error-prone rates in RNA viruses are approximately 10,000 (1×10^4) times greater than those found in DNA viruses (i.e., herpesviruses, smallpox), whose proof-reading apparatus removes aberrant viral DNAs during DNA replication. Thus the consequences for evolution, selection, and biology of RNA viruses are considerable. RNA virus populations never represent a homogeneous clone but instead embody a swarm of related RNA sequences clustered around a master sequence. This swarm is termed "quasispecies" and provides a fertile source of genetic variants that can respond to selective pressures such as infecting a resistant host. As a result, parts of the virus' genetic composition can change for its advantage. This process includes continuous replication, advancement, and spread. Thus RNA viruses like Ebola can evolve up to one million times faster than DNA viruses.

The high error rate of RNA viruses places a restriction on their genomic size, that is, the number of genes carried by the virus. Various RNA viruses carry from 4 to 10 genes; by comparison, DNA viruses (like herpesviruses or smallpox virus) carry hundreds of genes. DNA viruses, while requiring only a relatively few genes for their replication, carry a suitcase of many genes to provide the virus with a selective advantage. This suitcase contains accessory genes, not vital to the virus replication but important for enhancing the virus' survival and production of progeny. Hence, RNA viruses with far fewer genes must do as much as DNA viruses that contain a multitude of genes. RNA viruses accomplish that task in part by encoding proteins that perform multiple tasks. For RNA viruses, this enhanced diversity leads to numerous individual progeny and loss of many viruses from the swarm due to lethal virus mutation. The advantage for RNA viruses like Ebola and Lassa is a rapid evolutionary response.

Finally, RNA viruses are further divided into positive- and negative-strand varieties. Viruses with a positive-strand RNA deliver their genomic RNA directly to cells' machinery (ribosomes) to begin their infectious cycle. Positive-strand messenger RNA (mRNA) viruses are infectious and include those like poliomyelitis and Coxsackie. By contrast, Ebola and Lassa are negative-strand RNA viruses. Their RNA is not infectious. These viruses must begin their infectious cycle by transcribing (copying) viral mRNAs. This reaction is catalyzed by enzyme(s) carried into the hosts' cells by the infecting virus.

The natural reservoir of infection for Ebola virus is unclear. Monkeys, other bush animals and fruit bats can be infected by the virus and serve as intermediate hosts. Infected monkeys spread the disease to humans when they enter the human food chain as edible meat or by blood contamination during butchering. Bats are believed to spread the disease by both their saliva, which then infects the fruit they suck/eat, parts of which are recovered and used in drinks consumed by humans and by their use as a source of food.[14,15] Thus man is an interloper who accidentally comes in contact with fluids or tissues from infected monkeys and bats. Once humans are infected, the virus replicates rapidly and is excreted in body fluids like tears, sweat from skin, blood, vomit, or diarrhea. It is the human-to-human spread of disease that causes epidemics. Not until the fundamental public health approach of rapid diagnosis, quarantine, which prevents contact between sick humans and their healthy counterparts, and the rapid burials of the stricken dead is the spread of disease halted. To restrict disease spread, susceptible persons must be removed from all sources of infection.[16]

In Central Africa, the migration of persons from village to village is limited by dense jungles, lack of easily accessible travel routes, and few vehicles for transport on land or water. Thus outbreaks of disease usually remain within one village and do not necessarily penetrate surrounding communities. By experience in Central Africa, once Ebola appears in a village, locals often isolate sick persons to a single place so that person-to-person contact is avoided. Food and liquid are left outside the isolation hut or house, which is then destroyed by fire after all the infected persons have died.

In contrast, West Africa has more accessible paths, roads, and means of travel that facilitate human migration throughout one area and across/back-and-forth borders of several countries. Further, since only one single case and no major outbreaks or infection of multiple individuals with Ebola occurred in West Africa before the recent epidemic, the experience of health care workers was limited and the disease itself was not considered a top priority in this part of Africa. As noted above, that single incident of a human infected with Ebola virus was located in the Ivory Coast (1994) transmitted during an autopsy on a diseased chimpanzee in the Taï Forest. The sequence identity between Ebola-Zaire of the Congo and the Taï-Ebola virus was only about 65%. However, the relatively inexperienced West Africans lacked public encounters with Ebola or memory of its devastating consequences; therefore planning for what to do and how to monitor the local population were virtually nonexistent. Therefore controls to limit person-to-person contact, discontinue hazardous tribal burial traditions, foster quarantines, and deal with the fear of viral spread were not yet sufficiently available to educate the susceptible population.

REFERENCES

1. Peter Piot and the Ebola Outbreak in the Yambuku in 1976. LSHTM Library; 2013. http://lshtmlib.blogspot.com/2013/10/peter-piot-and-ebola-outbreak-in.html [accessed 02.04.16].

2. Oldstone MBA. Hantavirus. *Viruses plagues & history*. New York, NY: Oxford University Press; 2010. p. 221−26.

3. Oldstone MBA. *Viruses, plagues, & history*. New York, NY: Oxford University Press; 2010.

4. Andersen K, et al. *Clinical sequencing uncovers origin and evolution during the seven months in Sierra Leone* 2015;**162**:738−50.

5. Sanchez A, Geisbert TW and Feldmann H. Filoviridae: marburg and Ebola viruses. *Fields virology*. 4th ed. Philadelphia, PA: Lippincot Williams & Wilkins; 2001, p. 1297−9.

6. CDC. Outbreaks chronology: Ebola virus disease. Centers for Disease Control and Prevention; 2015. http://www.cdc.gov/vhf/ebola/outbreaks/history/chronology.html [accessed 04.03.15].

7. Tai Forest National Park-Cote D'ivoire. African Natural Heritage; 2016. http://www.african-naturalheritage.org/tai-forest-national-park-cote-divoire/ [accessed 10.03.16].

8. Report of an International Commission. Ebola haemorrhagc fever in Zaire, 1976. *Bull World Health Organ* 1978;**56**(2):271−93.

9. Heymann DL. Ebola: Learn from the past. Nature.com; 2014. http://www.nature.com/news/ebola-learn-from-the-past-1.16117 [accessed 15.06.16].

10. Heymann DL, et al. Ebola hemorrhagic fever: Tandala, Zaire, 1977–78. *J Infect Dis* 1990;**142**(3):372–6.

11. Verger R. Newsweek rewind: The Ebola outbreak of 1995. Newsweek; 2014. http://www. newsweek.com/newsweek-rewind-ebola-outbreak-1995-262234 [accessed 18.03.16].

12. Levitt AM. Case history: Ebola hemorrhagic fever in zaire. Environmental Toxicology and Human Health II; 1995.

13. Hildebrandt A. Ebola outbreak: why Liberia's quarantine in west point slum will fail. CBC News World; 2014. http://www.cbc.ca/news/world/ebola-outbreak-why-liberia-s-quarantine-in-west-point-slum-will-fail-1.2744292 [accessed 06.03.16].

14. Callaway E. Hunt for Ebola's wild hideout takes off as epidemic wanes. Nature.com; 2016. http://www.nature.com/news/hunt-for-ebola-s-wild-hideout-takes-off-as-epidemic-wanes-1.19149 [accessed 16.05.16].

15. Leroy EM, Kumulungui B, Pourrut X, Rouquet P, Hassanin A, Yaba P, et al. Fruit bats as reservoirs of Ebola virus. *Nature* 2005;**438**:575–6. http://dx.doi.org/10.1038/438575a. [accessed 15.06.16].

16. Ebola (Ebola Virus Disease). Centers for Disease Control and Prevention; 2016. http://www. cdc.gov/vhf/ebola/about.html [accessed 07.03.16].

CHAPTER 2

Ebola's Unanticipated Arrival in West Africa

The Ebola epidemic of 2013−15 duly arrived in Guinea, West Africa, at the remote village of Meliandou in the district of Guéckédou, which bordered two other West African countries, Liberia and Sierra Leone.[1] At that time, December 2013, Meliandou was a rural forest community whose occupants farmed and hunted for the food supply to serve their families in just 31 huts. Chimpanzees and fruit bats were known sources of meat for the village and both can be infected by Ebola virus.[2] How the Ebola-Zaire virus, which first appeared a thousand miles away (2371 airline miles) in Yambuku at the heart of Central Africa's Congo 37 years earlier (1976),[3] now reared its head in Meliandou, West Africa, remains a mystery.

Emile Ouamouno, a 2-year-old boy, was known to have been exposed to and eaten both chimpanzees and fruit bats as food. Villagers also reported a tall, scarred, and hollow tree at the edge of Meliandou that housed a large colony of bats they called *lolibelo*. The tree was about 160 ft from the Ouamouno's hut and close to a small river used for washing. Local children, including Emile, frequently played in the hollow tree. Nearby, bats also hung under roofs of buildings.

In early December, Emile developed fever, vomiting, and passage of a black stool. Although his excretions were not examined for hemoglobin, a tarry or black stool is suspicious for oozing of blood or bleeding in the intestine. Four days later Emile was dead. From the onset of his disease on December 2, 2013, through March 26, 2014, 11 neighbors died in his village (nine deaths from December 2, 2013 to February 8, 2014; two deaths on March 26, 2014). First, Emile's mother died on December 13, his 3-year-old sister became sick on the December 25 and died on December 29. His grandmother died January 1, 2014; a nurse became ill on January 29 and died February 2, 2014. Finally, a village midwife became sick, was taken to the hospital at Guéckédou on January 25, and died on February 2, 2014.[1]

Ebola's Curse. DOI: http://dx.doi.org/10.1016/B978-0-12-813888-5.00002-0

Soon, in nearby villages like Dandou Pombo and Dawa, and then in medical centers including the Guéckédou Hospital and other neighboring health care sites and hospitals, similar deaths followed, all victims of *Ebola's curse*. Dr. Kalissa N'Fansoumane of Guéckédou remarked "…we thought it a mysterious disease." For example, from the neighboring village of Dawa, the sister of Emile's grandmother and a friend, both attended the grandmother's funeral in Meliandou. They followed the local custom of touching and washing the corpse's body. Upon returning home to Dawa, in late January both became ill with fever, diarrhea, vomiting, and bleeding, and they died before the end of January 2014. Eight such deaths were recorded in Dawa from January 26 to March 27, 2014. As for the adjoining village of Dandou Pombo, there were six deaths from February 11 until March 31, 2014. The disease was introduced there by a midwife from the Meliandou village who became ill after being cared for by a family member at the village of Dandou Pombo. On and on, the virus traveled from village to village to village. In the district's Guéckédou Hospital a health care worker passed the sickness on to others there, and those who became infected carried Ebola to the Macenta Hospital. Patients who died at Macenta Hospital had funerals in the village of Kissidougou. As a consequence of the funerals, five deaths followed on March 26, 2014, Ebola virus had been transported from Macenta to Kissidougou (Fig. 2.1).[4]

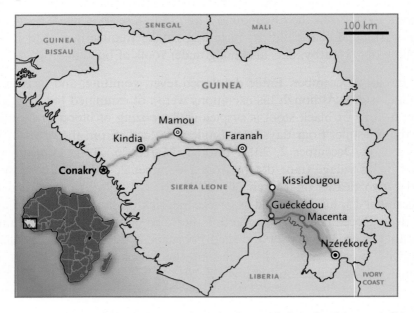

Figure 2.1 Map of the area of the index case, initial spread in Guinea and relationship of that area in Guinea to Sierra Leone and Liberia.

After an inconceivable lag period of over 3 months since the first deaths, on March 10, hospitals and public health services in Guéckédou and Macenta finally alerted the Ministry of Health in Guinea of their attack by an unknown "mysterious" fatal disease. Two days later, doctors without borders (Médecins Sans Frontières) were alerted about clusters of a mysterious disease characterized by high fever, diarrhea, vomiting, and a high fatality rate. Also in March the World Health Organization (WHO) was notified of the outbreak from an unknown pathogen that was devastating the region and rapidly spreading. The worrisome aftermath was that bureaucratic neglect had allowed a delay of over 100 days (109 days) after the first death occurred. Only on March 23, 2014 was the WHO to declare that the viral disease in West Africa was caused by Ebola.[5]

An investigative team from the Guinea Ministry of Health reached the outbreak areas by March 14; the doctors without borders members arrived on March 18. They undertook epidemiologic investigations and rapidly requested help to identify the etiologic agent responsible for the mysterious disease. Blood samples were collected and sent to the high-level BSL/4 containment laboratories in Lyon, France, and Hamburg, Germany. In those laboratories, RNA was extracted from 50 to 100 λ of plasma (a teaspoon contains 500 λ), diluted in RNA amplification reagent and searched for novel viral sequences that would be recognized as different from other infectious agents and as markers of the patient's affliction. An additional 100 λ of serum (blood fluid after removal of blood cells) was mixed with cultured cells to seek the expression and growth of a foreign agent. Cells used for this process were Vero E6, and supernatants from the cultures were repeatedly passed through the cells (a process to amplify the infectious agent). To identify the virus, antibody to several suspected infectious agents was applied in the hope of producing a reaction. Since the two most prevalent African hemorrhagic fever viruses are Lassa (common in that area of West Africa) and Ebola (except for one case in 1994 yet to appear in West Africa), these infectious agents were the major suspects. The results were unexpected and astonishing. The Pasteur group, led by Sylvain Baize, discovered that Ebola was involved in the outbreak and the Zaire strain was the specific subtype.[6] Of the blood samples collected from 20 patients, 15 were positive for Ebola virus. In viruses recovered from the cultured cells and subjected to electron microscopy, 1 of 2 samples tested showed a portrait characteristic of

Ebola (Fig. 1.1).[7] None of these samples according to any of the assays used was positive for Lassa fever virus or other suspected infectious agents (typhoid, malaria, yellow fever, etc.).[6] Thus, for the first time, Ebola virus infections of epidemic proportions were present and spreading in West Africa. The Pasteur group, in response to the French Ministry of Foreign Affairs, established a diagnostic laboratory in Macenta, Guinea. The report of Ebola in Guinea was rapidly placed on the internet on March 23, 2014. Those data were reviewed by members of the Viral Hemorrhagic Consortium at the Broad Institute, Boston, particularly two participants of that group, Kristian Andersen and Stephen Gire. Andersen and Gire left Boston for Sierra Leone and arrived at the Kenema Government Hospital (KGH), where they used the Lassa fever ward laboratory to set up a diagnostic polymerase chain reaction (PCR) assay for the specific detection of Ebola. Although at this time, no cases of Ebola had emerged in Sierra Leone, the investigators reasoned that the nearby open border with Guinea and the continuous exchange of travelers through both country's open borders would inevitably invite Ebola to pay a visit. Unfortunately, this prediction came to fruition within a short time.

The laboratory results in Sierra Leone and elsewhere yielded complete sequencing and bioinformatic data. Samples of Ebola virus circulating in infected patients in West Africa showed over 96% (96.8%) identity to Ebola-Zaire strains from the Democratic Republic of the Congo in Central Africa.[8]

The epidemiologic survey officially assigned the outbreak to Emile Ouamouno, the 2-year-old boy who was also the first one to die of the disease. Some of the food Emile was known to eat came from local animals (chimpanzee, fruit bats), previously established reservoirs for Ebola virus. A background check noted that Emile played in a partly hollowed-out tree where bats resided. Epidemiologists learned that the hollow tree in question caught fire and burned in late March, a few months after Emile died, and "...a rain of bats came out of the tree."

Researchers collected and tested bats in the area and found that they carried fingerprints of the Ebola-Zaire strain.[9] However, infectious Ebola virus has yet to be isolated from bats. Other scientists at the Robert Koch Institute in Berlin, including ecologists, veterinarians, and anthropologists later surveyed wildlife forests near Meliandou.[10] They found no evidence of a die-off among larger animals such as monkeys, which can be susceptible to Ebola, and suggested that perhaps some other yet to be identified animal might be responsible for the transmission to humans.[10] Fabien Leendertz, who led the study and questioned local residents stated "...We spent eight days in Meliandou. They told us that they regularly catch bats, like every other village in Guinea, Sierra Leone, and Liberia. The evidence is not 100%, and we can only say that this is the possible source of transmission. I would not say that the virus has emerged from central Africa. However, there are huge colonies of bats which regularly migrate. They can travel far in one night. I don't think an individual bat or colony migrated all the way from Congo or Gabon to West Africa. These big colonies are connected. There is a possibility for the virus to mix between colonies. They (the bats) share the same fruit. It is likely not to have even been one species of bat. The virus may jump from one species to another."[11] The team also collected samples of blood and tissue from captured bats, and the resulting data revealed their content of Ebola viruses, suggesting that the reservoir host was most likely bats. However, as of this writing, Ebola has not yet been isolated in the bats of the Meliandou area.

Like chimpanzees, bats carry the virus and could transmit Ebola via a soup made from bats considered a delicacy in Guinea. During the Ebola outbreak, Guinea banned the consumption and sale of bats for food in hope of preventing further spread of infection. Regardless, many natives in rural areas continued to hunt, procure, and eat bats and other bush meat. This practice proved to be a continual difficulty for controlling the spread and transmission of disease. Also, failing to convince the population to stop cultural practices like touching and bathing sick persons and the dead enabled the high viral titers in skin, tears, and saliva to continue their deadly onslaught by passage to new susceptible individuals.[12]

Burials have been a prime source of spreading Ebola virus infection. Common religious rites of family and friends of infected cadavers has created an overwhelming problem in containing the spread of the virus. In West Africa many cultural beliefs and religious rites after death includes washing, kissing, and touching the bodies of the deceased with bare hands.[12]

Yet, the optimal safety procedure to protect the living from becoming infected with Ebola is to remove the dead body while the burial crews wear full protective gear, seal the body in a bag, drench the bag in bleach, and place it in a grave 6-feet deep and fully covered. Unsafe burials played a substantial role in the spread of the virus. According to the WHO, "At least 20% of new Ebola infections occur during burials of deceased Ebola patients."[12] However, frequent push back and agitation between families and health workers interfered.[13] Sheku Bockaire, a Kenema Hospital health officer explained that "...putting people in body bags creates a lot of suspicion in the minds of people; they think parts of the body are being cut, and that's why the body is not being allowed to be displayed."[14] One of the WHO's top Ebola experts, Pierre Formenty, recommended stronger cultural sensitivity. "By building trust and respect between burial teams, bereaved families and religious groups, we are building trust and safety in the response itself. Introducing components such as inviting the family to be involved in digging the grave and offering options for dry ablution and shrouding will make a significant difference in curbing Ebola transmission."[12]

Death rituals in West Africa are rooted in a mixture of cultural and indigenous traditions as well as religious beliefs (predominantly Muslim and some Christian). According to customary ritual the deceased must be buried correctly otherwise his/her spirit will haunt the living and cause harm. Such rituals and fears ingrained in many years of tradition and passed down through generations explain why families of deceased Ebola patients are both distraught and angered when told they cannot bury that person "correctly."[15] Burials are seen as necessary to put the deceased's spirit to rest so that the ancestor does not remain to haunt the living. Death rituals begin by preparing the home of the deceased for mourning. Mirrors and pictures of the dead individuals are removed. Although these are not items for spreading the virus, removing the deceased's bedding is dangerous, since viruses are transferred in bodily fluids embedded in fabrics. Similarly, the rite of removing clothes from the deceased before wrapping the corpse in a sheet is a major problem, again because large amounts of infectious virus are found in sweat, tears, and bodily fluids of clothing from the deceased.

Because the rituals behind funeral practices are strongly held, difficulties arose in Liberia, after communities ran out of proper burial sites, and the government ruled that bodies were to be cremated. As a result, many people avoided hospitals, attempted to remove deceased family members and friends from hospitals, and conducted secret burials. In several instances health care workers were bribed to hand bodies over to families.[16] Desperate to avoid cremating their loved ones, many families began paying private retrieval teams to help hide the infected bodies from the hospital. These private groups issued certificates from the Ministry of Health stating that the patient is free from Ebola and thus can be released back to the family; prices of certificates range from $40 to $150. In addition, health care workers were being paid to produce fake death certificates making it difficult to track where bodies were buried. Some of those who could not hire these retrieval teams resorted to strikes and violence, causing workers to demand higher "risk" pay in fear of being attacked.[17]

Ebola was primed to spread to the neighboring country of Sierra Leone from where it was first found in the village of Meliandou. Sierra Leone shares a border close to Meliandou and the Guéckédou district of West Africa (Table 1.1) where young Emile Ouamouno died from Ebola virus infection in December 2013.

The first incidence of Ebola virus infection reached Sierra Leone on May 23, 2014. A young woman, Yillah, was admitted to the KGH after having a miscarriage.[18] Hospital staff workers were alerted and suspicious of Ebola infection due to its rapid spread, number of cases occurring in nearby Guinea, and the clinical presentation of the patient. A blood test using the PCR protocol set up earlier by Kristian Andersen and Stephen Gire confirmed that the young woman was positive for Ebola virus, and she was placed in an isolation medical unit. At that time there were 14 isolation beds available at KGH. Following her admission and treatment, no patients or staff at KGH contracted Ebola. The patient made a full recovery and was discharged. Afterward, health care workers and doctors at KGH began tracing back the source of her infection and its location. Their investigation determined that the patient had contracted the virus at the burial of a traditional native healer.[4] That healer was a female shaman and well-known for her mystical powers in and around the area of Kenema in Sierra Leone and bordering Guinea. Finda Nyuma's reputation not only included medicines she made from herbs in the forest, but also as people in and around her village area believed she communicated with the dead. Many came to her with messages to give to their dead family and friends. "They often found her beneath a bamboo palm, reading the future by throwing 'jagay', small white cowrie shells."[4] As the Ebola outbreak in Guinea grew relentlessly and uncontrolled, patients from Guinea traveled to the healer in Sierra Leone in search of treatment. Eventually, the famous healer died from Ebola infection. Soon, Finda Nyuma's relatives, neighbors, and friends came to her room to say farewell and prepare her for the afterlife. Most believed that she would then become an ancestor (spirit) who, in exchange for tribute and respect, could intervene with spirits on their behalf. Later, hundreds of mourners from many villages traveled to congregate at her funeral and subsequent burial. There the cultural practice of touching, washing, and lying on the body was again followed. As a result, 13 people at the funeral became directly infected. From these 13 the virus spread to over 300 other individuals resulting in as many as 365 deaths. Ebola was on the march to Sierra Leone. That march would result in Ebola-sickened victims fleeing to KGH which was approximately 63 miles from the healer's burial site in Koindu. The isolation ward at KGH with only 14 sick-beds soon housed as many as 80 Ebola virus-infected patients lying on floors and in halls. The hospital staff soon became overwhelmed, working close to 18 h daily and was to lose its head medical doctor Humarr Khan, head nurse Mbalu Fonnie, and nearly 50% of its staff.

REFERENCES

1. Ground zero in Guinea: the Ebola outbreak smoulders—undetected—for more than 3 months. WHO; 2016. http://www.who.int/csr/disease/ebola/ebola-6-months/guinea/en/ [accessed 12.05.16].

2. Quammen D. Insect-eating bat may be origin of Ebola outbreak, new study S. National Geographic; 2014. http://news.nationalgeographic.com/news/2014/12/141230-ebola-virus-origin-insect-bats-meliandou-reservoir-host/ [accessed 06.03.16].

3. Peter piot and the Ebola outbreak in the Yambuku in 1976. LSHTM Library; 2013. http://lshtmlib.blogspot.com/2013/10/peter-piot-and-ebola-outbreak-in.html [accessed 02.04.16].

4. Sack K, Fink S, Belluck P, Nossiter A. How Ebola roared back. NY Times; 2014. http://www.nytimes.com/2014/12/30/health/how-ebola-roared-back.html [accessed 02.03.16].

5. Ebola virus disease in Guinea. WHO; 2014. http://www.afro.who.int/en/clusters-a-programmes/dpc/epidemic-a-pandemic-alert-and-response/outbreak-news/4063-ebola-hemorrhagic-fever-in-guinea.html [accessed 16.04.16].

6. Baize S, et al. Emergence of Zaire Ebola virus disease in Guinea. *N Engl J Med* 2014;**371**:1418−25 http://dx.doi.org/10.1056/NEJMoa1404505 [accessed 03.05.16].

7. Geisbert TW, Pushko P, Anderson K, Smith J, Davis KJ, Jahrling PB. Evaluation in non-human primates of vaccines. *Emerg Infect Dis* 2002;**8**(5):503−7 http://www.ncbi.nlm.nih.gov/pubmed/11996686 [accessed 11.06.16].

8. Maganga GD, et al. Ebola Virus Disease in the Democratic Republic of Congo, N Engl J Med. 2014; 371:2083-91. DOI PubMed.

9. Reuters T. Bats likely carry Ebola to humans, but may also carry cure – technology & science – CBC News. CBCnews. November 03, 2014. http://www.cbc.ca/news/technology/bats-likely-carry-ebola-to-humans-but-may-also-carry-cure-1.2821851 [accessed 15.04.16].

10. Vidal J. Ebola: research team says migrating fruit bats responsible for outbreak. The Guardian. August 23, 2014. https://www.theguardian.com/society/2014/aug/23/ebola-outbreak-blamed-on-fruit-bats-africa [accessed 10.06.16].

11. Vogel G. Bat-filled tree may have been ground zero for the Ebola epidemic. Science and AAAS; 2014. http://www.sciencemag.org/news/2014/12/bat-filled-tree-may-have-been-ground-zero-ebola-epidemic [accessed 09.03.16].

12. New WHO safe and dignified burial protocol-key to reducing Ebola transmission. World Health Organization; 2014. http://www.who.int/mediacentre/news/notes/2014/ebola-burial-protocol/en/ [accessed 10.03.16].

13. Schnirring L. Probe of Ebola burial practices pinpoints risks, triggers change. CIDRAP; 2015. http://www.cidrap.umn.edu/news-perspective/2015/01/probe-ebola-burial-practices-pinpoints-risks-triggers-changes [accessed 15.03.16].

14. Fletcher P. As Ebola stalks West Africa, medics fight mistrust, hostility. Reuters. July 13, 2014. http://www.reuters.com/article/health-ebola-westafrica-idUSL6N0PO0V220140713 [accessed 15.03.16].

15. The editors of encyclopaedia britannica. African Religions. Encyclopaedia Britannica. http://www.britannica.com/topic/African-religions [accessed 22.04.16].

16. Ebola cremation ruling prompts secret burials in Liberia. Th Guardian. October 24, 2014. http://www.theguardian.com/world/2014/oct/24/ebola-cremation-ruling-secret-burials-liberia [accessed 22.04.16].

17. Charlton C. Bribery breaks out in battle against Ebola: Liberian Victims' families paying corrupt retrieval teams to keep bodies so they can give them traditional burials. Daily Mail. October 14, 2014. http://www.dailymail.co.uk/news/article-2791911/bribery-breaks-battle-against-ebola-liberian-victims-families-paying-corrupt-retrieval-teams-bodies-traditional-burials.html [accessed 29.04.16].

18. Sierra Leone: a traditional healer and a funeral. World Health Organization; 2016. http://www.who.int/csr/disease/ebola/ebola-6-months/sierra-leone/en/ [accessed 12.04.16].

CHAPTER 3

Kenema Government Hospital: From Lassa to Ebola

The Kenema Government Hospital (KGH), located 186 miles from the Sierra Leone capital of Freetown, is a large public health facility in this Western African country. Interest in the city of Kenema and the population surge there correlated with the discovery of diamond mines and building of a railroad in the 1930s. Kenema, now a major trade center with a population of just under 200,000, is the third largest city in Sierra Leone, after Freetown and Bo, and is the largest city in the Eastern province. For many years, KGH has provided medical service to the city of Kenema and surrounding areas. Currently, the hospital has approximately 200 beds. During 2003, after the end of the civil war in Sierra Leone, KGH established an advanced clinical and research center to manage Lassa fever, the major virus-induced hemorrhagic fever that was and is endemic in this part of Africa. Initially the Lassa virus treatment center run by the Ministry of Health and Sanitation of Sierra Leone comprised physicians and health care workers from Sierra Leone as well as those from Tulane University Medical School in New Orleans, Louisiana. The Sierra Leone Ministry of Health and Sanitation and the Viral Hemorrhagic Fever Consortium established a research program and also worked in tandem to provide diagnosis, surveillance, investigation of cases, and tracking of contacts.[1] Since rodents are the natural vector for Lassa fever virus,[2] a plan for rodent control as well as community outreach and education followed. Funding for the Lassa research treatment program was/is provided in part by the United States National Institutes of Health (NIH), Centers for Disease Control (CDC), World Health Organization (WHO), Merlin Group, and Ministry of Health and Sanitation of the Sierra Leone government. Although the early Lassa treatment center at KGH had only 14 beds, this facility came to play a pivotal role in treatment for victims of this major African hemorrhagic fever virus. Unexpectedly, KGH was also destined to become a crucial participant in the oncoming Ebola virus onslaught of 2014–16.

Ebola's Curse. DOI: http://dx.doi.org/10.1016/B978-0-12-813888-5.00003-2

Lassa virus was and continues to be the major cause of hemorrhagic fever in Western Africa, infecting roughly 400,000–500,000 individuals per year and yielding a death rate of 30,000–40,000.[2-4] Areas of prevalence for Lassa virus infection extend from Guinea in the West to Mali in the North to Nigeria in the East. However, the Eastern province of Sierra Leone has the distinction of harboring the highest incidence of Lassa virus-induced disease in the world. Infection from Lassa virus can occur all through the year but peaks in the rainy season, November to April. Typically, Lassa virus causes a persistent infection of rodents, and the human index case usually results from infection with virus shed by rodents in their urine or other excretions. During the rainy season rodents most often seek shelter inside village huts where contact with humans and stored foods is virtually unavoidable. In any 1 year, Lassa viruses infect and kill far more persons than the other major African hemorrhagic fever virus, Ebola. That is, Ebola, since the time it was first described to the present, including the major 2013–16 outbreak, has infected and killed fewer individuals in total than Lassa virus does in 1 year.[4,5] However, both Lassa and Ebola viruses cause devastating diseases with quite similar clinical features.[2,4]

Because of the high incidence of Lassa viral infection in Kenema, a Lassa fever isolation ward that is a center of excellence in the treatment and clinical research of this disease was built at KGH. KGH also serves as a base for treatment of the many other infections that occur in the area from malaria, yellow fever, and tuberculosis, to intestinal parasites as well as other medical issues.

In the early-1970s, Dr. Joseph McCormick of the CDC set-up a field site at the Nixon Memorial Hospital in Segbwema,[6] which was an investigative and treatment site at Ranguma 62 miles away from Kenema. This field station was established to investigate a then mysterious hemorrhagic fever first recognized in 1969 that affected five health care workers.[7] The initial report of this infectious disease by John Frame and colleagues in 1970, described the events recorded below.[4]

Ms. Laura Wine, a nurse working in the small mission hospital, Church of the Brethren, in Lassa, Nigeria, was in good health until about January 12, 1969, when she complained of a backache. On January 20 she reported a severe sore throat, but the physician who examined her found no signs to account for her discomfort. The next day, she complained that she could hardly swallow; she had several small ulcers in her throat and mouth, an oral temperature of 100°F, and bleeding from body orifices and hospital-induced needle puncture wounds. By January 24, she was suffering from sleepiness and some slurring of speech; late in the day she appeared increasingly drowsy. On January 25, she was flown to Bingham Memorial Hospital in Jos, Nigeria. She died on January 26 after several convulsions.

A 45-year-old staff nurse, Ms. Charlotte Shaw, at the Bingham Memorial Hospital in Jos, Nigeria, was on night call when Ms. Wine was admitted on January 25, 1969. Ms. Shaw had cut her finger earlier picking roses for another patient. As part of her nursing care, Ms. Shaw used a gauze dressing on that finger to clear secretions from the patient's mouth. Only afterward did she wash and apply antiseptic to the small cut on her finger. Nine days later Ms. Shaw had a chill with headache, severe back, and leg pains, and a mild sore throat, a clinical picture similar to that of Ms. Wine, who had died 8 days earlier. Over the next few days, Ms. Shaw had chills with fever to 102°−103°F, headache, and occasional nausea. Seven days after the onset of symptoms, a rash appeared on her face, neck, and arms and spread to her trunk and thighs. The rash appeared to be petechiae (small hemorrhages), and blood was oozing from several areas of her body. Her temperature was 104.8°F. By February 12, her face was swollen; she had shortness of breath, a rapid, weak pulse...became cyanotic (bluish)...had a drop in blood pressure. Nurse Shaw died on the 11th day of illness. Autopsy showed the presence of fluids in each pleural (chest) cavity and in the abdomen. Thus like Ebola virus infection, once established in a human, the infection afflicting these two nurses, spread to other susceptible humans.

A 52-year-old nurse, Ms. Lily Pinneo, working at the same Nigerian hospital, Bingham Memorial, had nursed both these patients and had assisted in autopsy of the second patient. She collected blood and tissue samples. On February 20, she too developed a temperature of 100°F, followed 2 days later by weakness, headache, and nausea. After another 3 days, she had a sore throat and petechiae and was admitted to the hospital. Since this was the third case in progression, the physician decided to send the patient to the United States for diagnosis and treatment. She was flown to Lagos, Nigeria, where she lay for 4 days in an isolation shed, and then to New York attended by a missionary nurse. She was admitted to Columbia University Presbyterian Hospital (New York City) and was placed in isolation with full precautions.

Pinneo continued to be acutely ill with a temperature of 101.2°F. The first night after admission, her temperature rose above 105°F. She became extremely weak during the next 6 days. Specimens from Ms. Pinneo were carried to the Rockefeller Foundation Arbovirus Laboratory at Yale University Medical School in New Haven, Connecticut, for study. "…the patient recovered strength slowly, became fever-free and was discharged from the hospital on the May 3, 1969."

One month later, an experienced and internationally known virologist, Dr. Jordi Cassals of the Yale University Arbovirus Research Laboratory, after working in New Haven with specimens from the patient, Ms. Pinneo, became ill and developed symptoms compatible with the acute mysterious virus infection.[4] During his slow convalescence, virus was isolated from his urine. The virus was studied and assigned the name Lassa, after the region in Nigeria where it was first isolated. A short time later, a technician from the Yale Arbovirus Laboratory, traveled to Pennsylvania to visit his family over the Thanksgiving holiday, became sick and died from Lassa virus infection. Consequently, the Yale Arbovirus Laboratory decided not to perform any more experiments with Lassa fever virus and shut down research and handling of this infectious agent. *The New York Times* and other publications reported the virus "too hot to handle."[4] With the presence of highly pathogenic infectious agents the construction of novel isolation units for the study of such agents were required and designed. Now, to safely protect scientists and allow work on these infectious agents, biosafety laboratory (BSL)-four units were constructed and opened. Several years later, with several BSL-4 laboratories available, such "hot viruses" as Lassa fever and Ebola could be safely handled. Twelve BSL-4 laboratories are now located in various areas of the United States and one in Canada. The BSL-4 laboratory at the University of Texas Medical School, Galveston, is the one where the former Yale Arbovirus Laboratory unit relocated after leaving New Haven.

Lassa, like Ebola, is an RNA negative-strand virus. Unlike Ebola, Lassa belongs not to the filovirus family but is an arenavirus family member.[2,4] The Lassa virus genome is segmented into two sections, each with an RNA segment encoding two genes (total of four genes, two on each segment); its size is roughly 10.7 kilobases, and as viewed by electron microscopy, appears as polymorphic sand-like round particles (Fig. 1.1). In contrast, Ebola is larger at 19kb, is a nonsegmented RNA virus, encodes seven proteins, and by electron microscopy is long, thin, and worm-like (Fig. 1.1). The host that carries Ebola in the wild is uncertain, possibly the fruit bat, whereas the reservoir that bears Lassa virus in nature is known to be a rodent. At least one way in which Ebola and Lassa are alike is that humans are unintentional intermediate hosts who initially become infected by exposure to a persistently infected animal. The virus is then passed from human-to-human as an end-stage infection, which if not cleared, kills the host. The disease following either infection is severe, since both viruses are among the most deadly infectious agents known to man. Frequently, either virus contaminates an unsuspecting or over-burdened hospital worker taking care of ill patients or disposing of the dead. The index case usually comes from a rural area.

As a response to the high concentration of Lassa fever infections in Sierra Leone, in the town of Segbwema, Joe McCormick and colleagues established the first CDC field station to treat and study these patients in the 1970s. However, a civil war in Sierra Leone in 1991 made this station too unsafe as a work area, and the unit was closed. With their activities in Sierra Leone disbanded the CDC moved the unit to Guinea where their work continued.

The bloody civil war broke out in 1991 when a rebel army, the revolutionary united front (RUF) led by Foday Saybana Sankoh formed as an antigovernment guerrilla warfare group and fought to overthrow the existing government of Joseph Momoh.[9] The RUF rebels began by attacking sites along the border of Liberia and Sierra Leone and quickly expanded to take over Eastern Sierra Leone. This guerrilla group gained prominence through murder, amputations, and child recruitment along with every form of fear and violence fostered by support from Charles Taylor's National Patriotic Front of Liberia (NPFL).[8] To finance the war the rebels took over the country's natural resources (diamonds). Thousands of children were kidnapped, drugged, and forced to become soldiers in the rebel army.[9] Major instability followed throughout Sierra Leone with over half of the country's 4.6 million people murdered or displaced. Towns and government facilities were destroyed. The civil war lasted 11 years, disintegrated the economy of the area, and left over 50,000 persons dead. The overall result was loss of government authority, along with a displaced, homeless, starving, and sick population. The seeds of that disaster were to play an important role later in the 2013–16 Ebola outbreak. Diseases such as Lassa fever continued unabated and added to the national crisis.

As a direct consequence of that civil war, one component of the Lassa virus research project in Sierra Leone was shifted to KGH. Dr. Aniru Conteh, a physician who had joined Joe McCormick and Sierra Leone's CDC team in 1979 and was trained in handling Lassa viruses, then moved to KGH where he became head of a Lassa investigation unit.[1] He and the team he had assembled wanted to continue their clinical and research work in Sierra Leone. Dr. Conteh devoted his life to treating patients with Lassa fever, becoming a leading world specialist in this infection.

KGH was left intact during the civil war and remained open. Despite the lack of security and resources, Dr. Conteh and his staff continued their work on Lassa fever. Over the 11 years of 1991–2002 the KGH staff treated thousands of patients, including the civilian native population, UN peacekeeping forces, and civil war fighters. Simultaneously, hundreds of individuals with Lassa fever were treated at the Lassa unit by Dr. Conteh. In March 2004, Dr. Conteh, while handling samples from a Lassa fever patient, sustained a needle stick, became infected with Lassa fever virus, and died 18 days later. Dr. S. Humarr Khan, then a recent graduate from the medical school in Freetown, was selected by the Sierra Leone government to succeed him as Director of the KGH Lassa unit.[10]

In 2004 the KGH Lassa unit established a partnership with Tulane University School of Medicine, New Orleans, Louisiana.[1] For some time, Tulane had been the principle partner with the Mano River Union Lassa Fever Network program, composed of the WHO and ministries of Sierra Leone, Guinea, and Liberia. This diverse group of organizations strove to develop national and regional prevention and control strategies for Lassa fever in Africa, as well as performing laboratory research to better understand the pathogenesis of this hemorrhagic disease. Dr. Robert (Bob) Garry, Professor of Microbiology at Tulane University Medical School, became the principle investigator (head) of the Viral Hemorrhagic Fever Consortium in 2003/2004 and remains in charge of building and maintaining this program to study Lassa fever virus infection. A Lassa fever isolation ward of 14 beds, the only such isolation ward in the world, was established at KGH and run by both Dr. Khan and a clinical physician from Tulane. They were responsible for patient care, directing clinical research after approval by the appropriate institutional review committees, and training of other health care workers. The Lassa fever ward was physically separated from the rest of KGH and had its own staff of approximately 40. Anyone in Sierra Leone suspected of having Lassa fever was sent to KGH where medical care was/is free. Those with a confirmed diagnosis of Lassa fever were then admitted to the Lassa fever isolation ward, where regulations were strictly enforced. That is, medical and health care personnel were required to wear protective gowns, gloves, masks, and face shields. In place was a Lassa fever program that, in addition to providing hospital isolation and treatment, included a clinical research wing and fostered both Lassa virus awareness and ecology teams. The awareness team of five persons worked on outreach efforts including case investigations, surveillance, and public education. The ecology team of four trapped rodents, collected samples, and investigated contacts, their locations and their travel routes to and from areas suspected of contamination. On occasion, members of individual teams worked with/on another team. Clinical research was performed only after approval by institutional committees set-up by KGH, the Sierra Leone government, funding agencies such as NIH, CDC, and participating American medical schools or institutions. A routine clinical Lassa laboratory was located on the grounds of KGH, and a separate BSL-2 suite was available for handling samples from patients with suspected Lassa fever. To operate this BSL-2 laboratory, staff with expertise and the necessary equipment could identify Lassa virus. The research and BSL-2 program were overseen by Augustine Goba of Sierra Leone and Robert Garry from Tulane University School of Medicine. Goba, director of the KGH Lassa laboratory, diagnosed the first case of Ebola virus infection in Sierra Leone using a PCR assay, and Garry was involved in the development of a rapid diagnostic test[11] to detect Ebola.[12] Pardis Sabeti and colleagues from the Massachusetts Institute of Technology (MIT) Broad Institute and the Harvard University Medical School were brought in to sequence viral isolates and part of the human genome.[13] Although work with the whole infectious virus requires a BSL-4 facility, tests using individual viral proteins can be done in a BSL-2 facility, and handling of infectious blood in endemic African countries is considered safe in a BSL-2 unit.

It was in this setting that the first Ebola-infected patient arrived at KGH in March 2014.[14,15] Rapidly, following the appearance of this virus-infected woman, first 10–20 then hundreds of sick successors appeared in March and June, all afflicted with a mysterious lethal infection soon diagnosed as Ebola. The research team knew that Lassa fever is spread on objects contaminated by infected rodents or by products contaminated with blood or tissues from infected patients and usually peaks in November through April, during the rainy season when rodents seek shelter in village homes. Now masses of ailing locals were arriving but not at the expected time when Lassa infections prevail. Soon the numbers of patients seeking medical help swamped KGH, changing the hospital's priority from Lassa virus to an Ebola treatment and clinical care center. Before 2014, Sierra Leone had no known case of Ebola infection, but in 2014 to mid-2015 there were over 8000[15] and up to March 2016 over 14,100. The outreach teams and surveillance programs available for the surrounding countryside soon became overwhelmed and inadequate. It was in this mix that Humarr Khan, his hospital staff, Bob Garry, Augustine Goba and his laboratory staff, and a young physician from Tulane, John Schieffelin, were to confront the rising tide and a continuous influx of sick and dying individuals infected with Ebola virus. Augustine Goba, for his work and dedication during the Ebola outbreak at the KGH Lassa unit, was awarded a Presidential Citation from the Sierra Leone president, Bai Koroma. In recognition of his diligent and dedicated service in the fight against Ebola, especially as a viral hemorrhagic laboratory scientist who diagnosed the first Ebola case in Sierra Leone and led the Ebola testing nationally until other Ebola diagnostic laboratories were established.[16] John Schieffelin, as a result of his experience, published a definitive report on the clinical presentation of Ebola infection.[17]

For Schieffelin and his colleagues the unexpectedly tremendous influx of Ebola patients provided a large trove of data to characterize clinical features of the disease.[17] First, he was able to confirm the observations of others regarding the clinical presentation of fever as the most common symptom (89% of patients had fever), weakness (66%), dizziness (60%), diarrhea (51%), abdominal pain (40%), vomiting (34%), with only 1% presenting with bleeding. Among the novel findings was surprising variability in how individuals responded to the virus; some had relatively mild cases and others went downhill quickly. These findings correlated with the Ebola viral load carried in their blood. Of those having 100,000 (1×10^5) copies of virus per millimeter of blood, 67% survived, but only 6% survived a viral content of 10,000,000 (1×10^7) copies/ml. What impact the innate immune response had (i.e., patients' production of type 1 interferon, other cytokines and chemokines, macrophages, natural killer cells, etc.) or the effect of adoptive immunity (virus-specific T cell responses, antibodies) remains unknown. However, these complex physiological systems are the subjects of active investigation in my laboratory (M.B.A. Oldstone) and those of Garry, Sabeti, and the KGH Lassa unit with NIH support. The results are vital to inform those who will generate efficient and effective vaccines and adoptive antiviral therapies.

Another issue raised from investigation of the Ebola outbreak is the possibility of one human being a super viral spreader.[18] That is, where exceptionally large groups of susceptible humans infected by one especially virulent human (a "super spreader") or did many much smaller groups of people become infected by similar contact with multiple infected individuals? How/if the answer to these questions apply and whether the answer might relate to differences in total virus content carried by an infected host are unknown. After amino acid sequencing of several Ebola virus genomes, we do know that the cause is probably *not* genetic variations (mutations) in the virus.[19,20]

With the increased numbers of Ebola-infected individuals in Guinea and the first cases from areas just a 3- to 4-h drive away, Garry departed New Orleans and quickly returned to KGH sensing an absolute need to be physically present to direct and participate in educating the staff and obtaining protective equipment for health care personnel before the Ebola infection reached Sierra Leone in large numbers and control of the infection was lost. As Garry later related, despite his frequent pleas very little help arrived from the international community and the outbreak really spun out of control. From the emergence of Ebola in 2013 to May 23, 2014, there were officially 258 diagnosed or probable cases of Ebola viral-induced disease, all in Guinea or Liberia. The WHO was only days from announcing that the outbreak ended when Augustine Goba diagnosed the first Ebola case in Sierra Leone. Subsequent investigations revealed that the outbreak had not been confined to Guinea and Liberia, instead the official numbers under-estimated the true number of cases of Ebola virus infections.[21]

Despite the growing numbers, some believed that the Ebola out-break would burn-out in as soon as weeks to months. In an interview on National Public Radio in the United States on June 18, Robert Garry disagreed and recounted the deaths in the vicinity of Daru, but his warning was depicted as needlessly alarming. Doctors without bor-ders was among the groups that also appealed for more international assistance at this time, but was likewise labeled alarmist. Because the internet is unreliable in that area of West Africa, Garry made frequent trips to KGH in Sierra Leone alternated with travel to the United States in attempts to raise funds to combat Ebola, Garry remarked. "In June 2014, [I] was one of the few saying this outbreak could take a spin for the worse and spin out of control. Unfortunately, international response was way too slow. International effort in public health (was needed) to stop disease."[21] The viral hemorrhagic fever consortium headed and run by Garry targeted research as the best way to under-stand the pathogenesis of Lassa fever infection and was not a public health agency. Actual public health solutions to contend with Ebola required many boots on the ground for diagnosis, isolation, education, large scale case tracking, and disposal of the dead.

Garry related that some help came but not enough. He depicted the spread of Ebola like a train speeding down the track. People running after it, but not enough people to actually get them to stop it. He warned that the Ebola outbreak was so much more massive than any-thing that anybody had ever been confronted with before. He stated "the prior outbreaks − you can look them all up (see Table 1.1) − have all been in middle Africa. And they've been in small villages that were very isolated. It's much easier to get those patients isolated. Middle Africa doesn't have the road structure, doesn't have the popu-lation density that is there in West Africa. People move around a lot; the roads are good in West Africa." Remember, there are diamonds there, and that means that there is a very interesting geography and lots of valuable minerals. So mining companies have built a lot of good roads, and there are a lot of people moving around. So that is a recipe for the virus to be able to spread quickly.[21]

At KGH, a staggering number of sick individuals kept coming. There, Khan and other doctors from Sierra Leone, Tulane, and elsewhere were on the front line. The case load of KGH increased beyond its capacity. Nurses from other parts of KGH were recruited but personal protective equipment was in short supply. By the middle of July 2014, several members of the nursing staff had been infected with Ebola virus. One of the nurses who became infected was pregnant. Four nurses, including the head nurse of the KGH Lassa ward, Mbalu Fonnie, who worked with the Lassa fever program for over 25 years, attempted to save the life of their nursing colleague by inducing a still-birth delivery, a procedure that offers a chance of survival for the pregnant mother.[22] Despite their best efforts, though, the pregnant nurse died; then each of the four treating nurses became infected with Ebola virus. All four died, including nurse Fonnie.

The Tulane program directed by Garry provided the opportunity for young infectious disease fellows and staff to rotate to KGH and become involved in global health. One such trainee was John Schieffelin. Schieffelin, an Assistant Professor of clinical medicine and pediatrics, having finished his residency training at the Louisiana State University Health Science Center in New Orleans and Tulane University in 2009, went to join the Viral Hemorrhagic Fever Consortium to pursue his interest in clinical infectious disease, antibody responses, and global health. He was sent to KGH through a WHO program. At the KGH prior to the Ebola outbreak, Schieffelin profiled Lassa virus-infected patients. Serendipitously, Schieffelin was there when Ebola broke out and during the massive influx of Ebola virus-infected individuals, joining the subsequent, much honored Director of KGH, Dr. Khan, where they worked in the designated Lassa virus infection ward. There were times when 80–90 patients occupied that facility with only 14 beds. Those care providers were stretched much past their capacity, working 16- to 18-h shifts. Schieffelin, as a result, saw a multitude of Ebola-infected patients and was able to provide one of the best and most thorough descriptions of their clinical manifestations and outcomes. The result was sadly different for Dr. Khan. Khan became infected with Ebola, becoming one of its victims who died after contracting the disease from infected patients. The loss of such a prominent leader and so many health care workers had a devastating effect on their colleagues and coworkers. As stated by Garry "These people are my colleagues and my friends. I've been working with them for 10 years. It's devastating to have lost these very valuable colleagues and people that I care about. That's irreplaceable.

Figure 3.1 Top: Kenema Government Hospital. Lower left: Holding area where individuals were assessed for Ebola virus infection and admission to KGH or discharge for those who were not infected. Lower right: Displays the burial area located outside the hospital.

Dr. Sheik Humarr Khan, who you may have read about, caught Ebola. He was a person I worked with for 10 years. His legacy will be — he will be missed. We'll carry on. We'll move forward. But this is a person who it is impossible to replace. Nurse Mbalu Fonnie, we've been working with her for 10 years, but she's been doing Lassa fever research for 30 years. You can't just replace that kind of experience and know-how and find somebody who knows these things overnight. So, yeah, it's going to be a rebuilding process when this is over."[21]

"You just do what you can. Obviously the people who are working with the patients and are the ones on the front line of the treatment and are trying to give care to these people who are infected with Ebola are at the greatest risk. In the laboratory environment, that's a more controlled environment. We can feel more secure in there. But there's always a chance that a mistake is made...and so, yeah, these are considerations. This is high-risk work."[21] (Fig. 3.1).

REFERENCES

1. Kenema Government Hospital. Viral Hemorrhagic Fever Consortium. http://vhfc.org/consortium/partners/kgh [accessed 17.06.16].

2. Buchmeier M, de la Torre JC, Peters CJ. Arenaviridae: the viruses and their replication. In: Knipe DM, Howley PM, editors. *Fields virology*. 5th edition Philadelphia: Lippincott, Williams & Wilkins; 2007.

3. Ogbu O, Ajuluchukwu E, Uneke CJ. Lassa fever in West African sub-region: an overview. *J Vector Borne Dis* 2007;**44**(1):1−11. PMID 17378212. "Lassa fever is endemic in West Africa".

4. Oldstone MBA. Lassa fever. *Viruses, plagues, and history*. New York: Oxford University Press; 2010.

5. 2014 Ebola outbreak in West Africa − case counts. Centers for Disease Control and Prevention; 2016. http://www.cdc.gov/vhf/ebola/outbreaks/2014-west-africa/case-counts.html.25 [accessed 18.06.16].

6. McCormick JB, King IJ, Webb PA, et al. Lassa fever. *New Engl J Med* 1986;**314**:20−6.

7. McCormick JB, Fisher-Hoch SP. Lassa fever in Arenaviruses I. The epidemiology, molecular and cell biology. Curr Top Microbiol Immunol 262. In: Oldstone MBA editor, Springer-Verlag, Berlin; 2002.

8. Jang SY. The causes of the Sierra Leone civil war. E-International Relations; 2012. http://www.e-ir.info/2012/10/25/the-causes-of-the-sierra-leone-civil-war-underlying-grievances-and-the-role-of-the-revolutionary-united-front/ [accessed 18.04.16].

9. Sierra Leone rebels forcefully recruit child soldiers. Human rights watch; 2000. https://www.hrw.org/news/2000/05/31/sierra-leone-rebels-forcefully-recruit-child-soldiers [accessed 12.04.16].

10. Profile: leading Ebola doctor Sheik Umar Khan. BBC News; July 30, 2014. http://www.bbc.com/news/world-africa-28560507 [accessed 18.04.16].

11. Ebola rapid test under development. Guardian Liberty Voice; March 27, 2014. http://guardianlv.com/2014/03/ebola-rapid-test-under-development/ [accessed 22.07.16].

12. Hayden EC. Ebola virus mutating rapidly as it spreads. Nature.com; August 28, 2014. http://www.nature.com/news/ebola-virus-mutating-rapidly-as-it-spreads-1.15777 [accessed 17.06.16].

13. Pardis Sabeti, Broad Institute Voices from the Frontlines of an Epidemic − The Broad Foundation Report 2015−16. The Broad Foundation Report 201516; 2016. http://broadfoundationreport.org/portfolio/stories/science/pardis-sabeti-broad-institute-voices-from-the-frontlines-of-an-epidemic/ [accessed 27.07.16].

14. Hayden EC. Infectious disease: Ebola's lost ward. Nature.com; September 24, 2014. http://www.nature.com/news/infectious-disease-ebola-s-lost-ward-1.15990 [accessed 18.04.16].

15. Ebola virus disease outbreak. World Health Organization; 2016. http://www.who.int/csr/disease/ebola/en/ [accessed 18.06.16].

16. Sierra Leonean government honors VHFC team members. Viral Hemorrhagic Fever Consortium; January 24, 2016. http://www.vhfc.org/media/news/sierra-leonean-government-honors-vhfc-team-members [accessed 12.04.16].

17. Schieffelin JS, et al. Clinical illness and outcomes in patients with Ebola in Sierra Leone. *New Engl J Med* 2014;**371**:2092−100. http://dx.doi.org/10.1056/NEJMoa1411680 [accessed 15.06.16].

18. Wong G, et al. MERS, SARS, and Ebola: the role of super-spreaders in infectious disease. *Cell Host Microbe* 2015;**18**:398−401.

19. Gire SK, et al. Genomic surveillance elucidates Ebola virus origin and transmission during the 2014 outbreak. *Science* 2014;**345**:1369−72.

20. Park DJ, et al. Ebola virus epidemiology, transmission and evolution during seven months in Sierra Leone. *Cell* 2015;**161**:1516−26.

21. Catalanello R. Q&A with tulane researcher on front line of Ebola outbreak. NOLA.com; August 28, 2014. http://www.nola.com/health/index.ssf/2014/08/qa_with_tulane_researcher_who.html [accessed 02.05.16].

22. Gilbert KL. Nurses, doctors save lives but lose their own − The United Methodist Church. The United Methodist Church; 2014. http://www.umc.org/news-and-media/nurses-doctors-save-lives-but-lose-their-own [accessed 17.06.16].

CHAPTER 4

Sheik Humarr Khan: Leading the Fight Against Ebola in Sierra Leone at Kenema Government Hospital

Sheik Humarr Khan was born in 1975 into a large family of nine brothers and sisters. Their parents and some of his siblings live in Lungi, a town not far from Freetown. Even in his youth, Khan dreamed of a medical career. Growing up he frequently addressed himself as "doctor" in the presence of family and friends. His dream became a reality when he was accepted to, schooled at, and graduated from the University of Sierra Leone College of Medicine and Allied Health Sciences. During that time, Khan became deeply interested in Africa's infectious diseases, particularly LASV, Ebola, malaria, tuberculosis, human immunodeficiency virus, and its disease, AIDS.

After graduating from medical school, Khan took a 1-year internship with focus on tropical medicine and infectious diseases. He was recruited as a medical officer by the Directorate of Disease Prevention and Control Ministry of Health and Sanitation, in Sierra Leone. After 2 years in this position and with the death of Dr. Aniru Conteh, the hospital's former Director, Khan applied for and was accepted for the open position as head of the KGH Lassa fever program. That few applicants wanted the position was no surprise. Work with a dangerous pathogen like Lassa fever virus requiring hours in the BSL facility was not popular. Further, many Sierra Leone physicians were hesitant to move away from the academic centers where they trained and leave better economic opportunities. But Khan saw the bigger picture of KGH as worthy and a worthwhile prospect for a successful future. His involvement as a consultant for the Mano River Union Lassa Fever Network brought him into a special scientific relationship with Bob Garry and the Tulane group and Pardis Sabeti and the Genomic Center at Harvard/MIT Broad Institute. As its Chief at the KGH Lassa Fever Virus facility, Khan ran that LASV program for almost a decade. His knowledge and professional experience in viral

Ebola's Curse. DOI: http://dx.doi.org/10.1016/B978-0-12-813888-5.00004-4

hemorrhagic fever diseases attracted the attention of the United Nations. He was then contacted by the United Nations Mission in Sierra Leone (UNAMSIL) as a physician consultant for Lassa fever in Sierra Leone. From 2005 to 2010, Khan working at KGH was the physician in charge of HIV and AIDS regional services as well as those for Lassa fever. From 2010 to 2013, Khan temporarily left his post at KGH to complete an internal medicine residency at the Korle Bu Teaching Hospital in Accra, Ghana. With the outbreak of Ebola, though, he completed his studies and returned to his position as physician in charge of the Lassa fever program.

Dr. Khan and the KGH team came to the fight against Ebola with boundless energy and passion. Among them was a background of many years' expertise in the diagnosis, care, and treatment of the major African hemorrhagic disease of West Africa, Lassa fever virus, a plague that annually infected the Sierra Leone population.[1,2] The wholly unanticipated outbreak of Ebola virus in Sierra Leone in 2014 quickly swamped KGH turning its Lassa fever ward into an Ebola center. Better prepared than most hospitals with a staff trained to wear protective garments and use their antiviral armamentarium to fight Lassa, they easily and quickly adjusted to Ebola. Thus they came to wage a strong and competent fight against Ebola infection with experience and expertise gained from many years of exposure to Lassa. With the massive influx of cases, Dr. Khan's team expanded to over 40 persons including scientists, physicians, and medical care workers from Kenema, Sierra Leone, Tulane, Harvard, MIT, others from West Africa, United States, Europe, and Asia (Fig. 4.1). Medical efforts were led primarily by Khan, assisted by Americans from Tulane like Bob Garry, followed by John Schieffelin, as well as Pardis Sabeti and her crew from Harvard and the MIT Broad Institute. While Khan and Schieffelin with head nurse Mbalu Fonnie organized and provided medical services, Garry along with Augustine Goba, managed the diagnostic and research laboratory. In addition, Garry collected patients' blood samples to be analyzed for understanding their immune responses; that is, where these responses fighting the infection or not? From these blood samples, they also obtained RNA to test the Ebola virus' genetic pattern for changes and mutations. Some samples were then sent to Pardis Sabeti, a trained physician and geneticist at the center for genomics, who was instrumental in studying infectious diseases. Funded by the World Bank and the NIH, she and colleagues originally

Figure 4.1 Left panel: The monument dedicated to health care workers of KGH who lost their lives while engaged in the fight against Ebola. Right panel: Shows several of the prominent workers at KGH who played a role in the Ebola outbreak in Kenema 2014–15. From top to bottom, left to right: Dr. S. Humarr Khan, Dr. Donald Grant, current Chief Physician of the Lassa ward at KGH (first row); Drs. Bob Garry, Pardis Sabeti (second row); Augustine Goba, and a protected worker in the laboratory (third row); health care workers and MIT geneticists, from the right are Simbiie Jalloh (VHFC Program Manager), Mambu Mamoh (laboratory technician), Dr. Khan, Augustine Goba (Laboratory Director), Mr. Stephen Gire and Dr. Kristian Andersen (fourth row). Fifth row displays protective outfits worn by health care workers.

set up the instrumentation, bioinformatic programs, and personnel to analyze the genomics of Lassa fever virus (LASV) in Nigeria and at KGH, but with the tidal wave of Ebola cases, they also took on the genomic study of Ebola.

Additionally and importantly, Garry shuttled multiple times between Africa and the United States, attempting to raise funds to purchase protective equipment for the health providers, supplies for patient care, and resources to keep the KGH isolation ward functional (Fig. 4.1). His main concern, though, was to alert the WHO and Americans at large of the current Ebola disaster's ever enlarging scale and its coming threat. Nevertheless, Khan, like many, feared the lethal virus. He stated "I am afraid for my life. I may say...health workers are prone to the disease because we are the first port of call for someone who is sickened by disease."[3] Khan's courage, work ethics, personality and charge to fight Ebola inspired not only the KGH staff but many of the international colleagues, collaborators, and friends he knew. His predecessor, Dr. Conteh, had pricked himself with a needle contaminated with blood from a woman who had Lassa fever, and he died 12 days later from the infection. At the time of Khan's appointment, he was all of 30 years of age and hardly tested in administration, running a service, or handling large influxes of Lassa virus infected persons. However, in retrospect, he turned out to be an excellent choice to fill the position. He learned quickly and earned the respect of the KGH Lassa medical care staff, Bob Garry and the Tulane group, as well as Pardis Sabeti and the Harvard/MIT Broad Institute participants; they and others who interacted with him voiced their high regard while later mourning his death.

With the Ebola outbreak, 14 beds in the KGH Lassa isolation ward were rapidly filled with sick and dying patients as were adjacent areas on the floors, halls, and outside grounds. The beds in the KGH Lassa ward were "cholera beds"—that is a mattress covered with plastic but punctured with a hole in the center. A bucket was placed under the hole to catch the diarrhea output. As the numbers of Ebola patients increased and Khan's stress magnified, he spoke for himself and his coworkers who were justified in their fear of dying from early and continual exposure to their sickened charges.

His family in Lungi was concerned and they asked him to consider leaving KGH and no longer risking his life. Khan's sister said "I told him not to go there" but he said "If I refuse to treat them, who would treat me?" As the virus rampaged through West Africa, the Lassa containment unit at KGH became overwhelmed and filled by patients harboring the Ebola virus infection—their number exceeding the available health staff, services, and beds. Khan became emotionally and physically depleted. Still, he stood bravely by his post even when the outbreak grew overwhelming and continued, endlessly providing treatment. Not only was he in charge of leading the hospital, he also treated hundreds of patients himself. Meanwhile, he trained senior doctors and medical staff on precautionary measures as well as how to assist and treat patients. He was constantly on the phone with government officials to aid in coordinating control efforts and to raise needed funds.

Alex Moigboi, who worked in the hospital with Khan, was the first of the health care staff to contract Ebola. He died soon afterward. Then head nurse, Mbalu Fonnie, who had worked at the KGH Lassa ward since it first opened in the 1990s, became infected with Ebola and died. Two other nurses, Fatima Kamara and Veronica Tucker, became infected. Many staff members at KGH became terrified, deserted the hospital/ward and their work, never to return. However, Khan remained in the Ebola ward with less and less support and fewer and fewer supplies. The health care system was collapsing under the strain of Ebola. Elsewhere, other international medical groups were stretched thin. One such group, Doctors Without Borders, was barely coping with Ebola patients in several health care units including Kailahun, a town about 65 miles from KGH.

Khan talked regularly with Pardis Sabeti—"we are all alone here," he told her. In addition to Bob Garry, Pardis and her colleagues were working hard to rush supplies and people to KGH. She told Khan "the most important thing is your safety." "Please take care of yourself." "People and help are coming." But it was not to be enough or in time. Khan told Pardis "I have to do everything I can to help these people." Khan was commanding and leading a battlefield charge, although many troops under his command were dead, dying, or fleeing.

In May 2014 the first patient with Ebola was seen, diagnosed, and treated at KGH, but about 4 months later, on 18 July, Khan did not feel well. He had a blood test taken on July 21 with results that confirmed he was infected with Ebola.[3–6] Because of the likely psychological impact, he did not want his remaining staff to see him develop the symptoms and signs of Ebola, thus further compromising their already low morale. So Khan arranged for transportation to take him to Kailahun, the Ebola care and treatment center run by Doctors Without Borders.

At the Kailahun health center, as elsewhere, therapy was limited to patient care and fluid replacement. There were no proven pharmaceutical drugs available to directly treat the infection. However, the Kailahun Ebola care center did possess a potential but untested medical therapy, a cocktail of antibodies called ZMapp. Antibodies are proteins synthesized by specialized lymphoid cells in response to stimulation by a specific antigen, in this case antigen(s) (proteins) of the Ebola virus. By virtue of the antibody's two binding parts, it can attach to the specific viral antigen that elicited it, inciting a series of events that neutralize or block replication of the virus from which the antibody was derived.[7] Antibodies may also attach and crosslink the virus, thereby lowering the numbers of virus particles present.[7] ZMapp was still an experimental antibody preparation and not yet tested in humans. However, 3 months before Khan became ill, ZMapp had been given to monkeys infected with a lethal challenge of Ebola virus and fully protected them from developing the fatal disease.[8] That protection saved the animals' lives even when the antibodies were given after the monkeys became sick, 5 days after receiving Ebola.

Doctors had observed that humans who survived previous infection with Ebola resisted subsequent reinfection by that virus. Also, although controversial, several reports insisted that blood (serum or plasma) taken from those previously infected with and immune to the virus might be helpful as a lifesaving therapy for patients sick with Ebola. Theoretically, such blood could contain antibodies to Ebola and, when administered as a so-called adoptive transfer (of antibody), provide immune therapy. The difficulty was this; no one knew whether adoptively transferred antibodies were actually protective—after all, some Ebola-infected patients do not die—so those who survive after receiving adoptive immune therapy might have survived without the antibody therapy. Further, some Ebola-infected persons given the antibody transfer die. Nevertheless, when confronted with a ruthless disease and limited therapeutic options, the use of antibody transfer was the rationale for ZMapp Biopharmaceutical Inc. in San Diego, in collaboration with the Public Health Agency of Canada's National Microbiology Laboratory in Winnipeg, Canada, manufactured the ZMapp product. To do so, they isolated antibodies to Ebola from antibody-producing cells of individuals surviving Ebola infection or from mice immunized with Ebola virus. Such antibodies were then manipulated genetically to produce a strongly neutralizing product to be used as a therapeutic reagent. To enhance the amount of antibody made the investigators used tobacco plants from a biotech Kentucky firm, Kentucky BioProcessing. By inserting the genes for antibody production into these tobacco plants, a therapeutic compound was produced that could potentially treat patients infected with Ebola virus, and eventually yield large quantities of that compound to be used as treatment for victims of the epidemic. However, before being used for humans, the therapeutic ZMapp would first have to go through testing in laboratory animals to show potency and then clinical trials in humans to show safety. At this point, ZMapp was found effective against Ebola virus in tissue culture and in animal experiments.[8] However, its toxicity or therapeutic potential was unknown and untested in humans. Also, insufficient data were available to know how the ZMapp reagent would hold up in a less than optimal environment. The ZMapp stored at the Doctors Without Borders center in Kailahun would be a test of the compound's survival properties in an environment where electrical power often failed, and the climate was very hot.

The government of Sierra Leone regarded Khan's deepening sickness as a national crisis and sent out international pleas for any drug or therapy that might help him. A series of international conference calls were made to officials from the WHO, CDC, NIH, US Army Medical Research Institute of Infectious Diseases, Public Health Agency of Canada, ZMapp, and Doctors Without Borders where treatment for Dr. Khan was debated. In the end the decision *not* to treat Khan was made on July 25 by Doctors Without Borders. Khan was not informed of the availability of ZMapp, and he died 4 days later on July 29 in his 39th year of life. Of course, no one knew if ZMapp would work or whether it was toxic. This dilemma posed an ethical problem for treating West Africans because of the risks involved and past medical treatments of Africans.[9] Nevertheless, one might argue, and we would, that Khan should have been told about ZMapp and given the option of receiving the therapy.

When Ebola killed Dr. Khan, he had treated over 100 patients bearing that infection. He had worked shifts of 16–18 h/day administering to their overwhelming sickness, despite the lack of supplies. Overworked and exhausted, he became an easy target for the virulent Ebola. Donald Grant was recruited by the Sierra Leone government to replace Dr. Khan after his death and handle his responsibilities.

To honor Khan's work, sacrifice, and his good life and premature death, Khan was recognized by Sierra Leone's President Ernest Bai Koroma and termed a "National Hero" of the country.[4] On the grounds of Kenema Government Hospital a memorial monument was dedicated to him and to those at KGH who lost their lives in the fight against Ebola (Fig. 4.1). The LASV ward being rebuilt into a 40 bed unit was named after him. As a tribute from the United States, the American Society for Microbiology established the Sheik Humarr Khan Lecture and Prize at its annual national meeting. The first such named lecture was presented by Heinz Feldmann of the NIH, National Institute of Allergy and Infectious Diseases, Rocky Mountain Laboratories located in Hamilton, Montana. Feldmann, who directs the BSL-4 laboratory there, had been working with Ebola viruses and continues to do so. He knew and interacted with Khan. Dr. Khan was posthumously awarded the 2015 Ed Nowakowski Senior Memorial Clinical Virology Award from the Pan American Society for Clinical Virology, which was accepted in his absence by his brother, Sahid Khan.

On July 31 there was a funeral for Dr. Sheik Humarr Khan in Kenema at KGH. In attendance were over 500 mourners including the Sierra Leone governing health ministers, health care workers, and scientists who knew and worked with him. Residents of the town who joined them included many he had cared for. Khan was buried on KGH grounds newly named for him. Before the Ebola outbreak, Khan had been scheduled to take a few months sabbatical with Pardis Sabeti in Boston to learn more genetics. Instead, he lay in a grave adjoining a memorial in his honor at KGH.[10]

His colleague *Dr. Pardis Sabeti with Bob Katsiaficas* composed this song for him and other health workers who succumbed to the deadly Ebola virus while serving their patients in Sierra Leone.

One Truth
Verse 3
A lifetime that we write
We laugh
We cry
We pray
We are love

we dream
we scream
we strive
our hunger will never die
I'm here in this fight, always

A lifetime for one for one truth
That I'm alive, And so are you
We are here, We are the proof
Yeah

A lifetime for one for one truth x 3

REFERENCES

1. Oldstone MBA. Lassa fever. *Viruses plagues & history*. New York, NY: Oxford University Press; 2010.

2. Buchmeier M, de la Torre JC, Peters CJ. Arenaviridae: The viruses and their replication. In: Knipe DM, Howley PM, editors. *Fields virology*. 5th edition Philadelphia: Lippincott, Williams & Wilkins; 2007.

3. Sanchez N. Sheik Umar Khan, Sierra Leone's Top Ebola Doc, catches the disease. NewsMax; July 24, 2014. http://www.newsmax.com/TheWire/sheik-umar-khan-ebola-doctor/2014/07/24/id/584621/ [accessed 24.05.16].

4. Head doctor fighting Ebola outbreak in Sierra Leone contracts the deadly virus. Reuters News; July 23, 2014. http://www.telegraph.co.uk/news/worldnews/africaandindianocean/sierraleone/10986310/Head-doctor-fighting-Ebola-outbreak-in-Sierra-Leone-contracts-the-deadly-virus.html [accessed 04.06.16].

5. Hammer J. My nurses are dead, and I don't know I'm already infected. Matter; January 12, 2015. https://medium.com/matter/did-sierra-leones-hero-doctor-have-to-die-1c1de004941e#.fcqm5t8a4 [accessed 15.04.16].

6. Dr. Sheik Humarr Khan. Viral hemorrhagic fever consortium. http://vhfc.org/consortium/people/humarr-khan [accessed 16.05.16].

7. Oldstone MBA. Virus neutralization and virus-induced immune complex disease. *Prog Med Virol* 1975;**19**:84−119.

8. Qui X, et al. Reversion of advanced Ebola virus disease in non-human primates with ZMapp. *Nature* 2014;**514**:47−53.

9. Omonzejele PF. Ethical challenges posed by the Ebola virus epidemic in West Africa. *J Bioethics Inq* 2014;**11**:417−20.

10. Wilson J. Ebola doctor in Sierra Leone dies. CNN; July 31, 2014. http://www.cnn.com/2014/07/29/health/ebola-doctor-dies/index.html [accessed April 15, 2016].

CHAPTER 5

ZMapp: The Ethics of Decision Making

The development of treatments for human viral diseases consists of two quite different approaches. The first is research at the molecular level to define the virus' life cycle in its quest to produce progeny virus and then screen for pharmacologic drugs that block reproductive stages of that life cycle. To gain knowledge of what self-components the virus requires to allow its survival and production of progeny while interacting with the host, enormous computerized libraries of known chemicals or synthesized compounds are searched. The goal is to identify molecules with the ability to block viral replication. Such assays require a robust readout and automation involving robotics to screen virtually millions of test molecules to identify just a few that show promise. Any molecule identified requires, first, confirmation and, then, analysis that it is not toxic by itself. Thereafter, chemists study and modify its structure to optimize the therapeutic index, that is, the relative safety of a dose or treatment as opposed to its potentially harmful effect.

Achieving the maximal advantage for an eventual therapeutic product requires vigorous chemical and biological testing to evaluate the stability, half-life, and best route of delivery—either oral, intravenous, or subcutaneous. Armed with this knowledge, the developers' next steps of discovery are to determine the pharmacologic molecule's effectiveness, usually first in cell cultures and then in animal models. For testing Ebola virus, animal models can include genetically modified mice, guinea pigs, or subhuman primates.

In the course of this search, hundreds of thousands of molecules are often screened before a therapeutic "hit" is uncovered. Unfortunately and frequently, the molecule or compound identified is disqualified because of its toxicity, insolubility, or delivery problems. Chemists then work to alter the structure of the selected compound to overcome such difficulties.

Ebola's Curse. DOI: http://dx.doi.org/10.1016/B978-0-12-813888-5.00005-6

The time involved can be and is usually long and the financial cost being great. Nevertheless, this approach has achieved amazing success over the last few years. Examples are discoveries of drugs that changed HIV/AIDS infection from a near-routinely lethal event (mortality >90%) to a 1% or less death rate for HIV patients who are medicated daily. However, these drugs are expensive and not always available in some countries. Similarly, the antiviral therapy recently discovered for chronic hepatitis C virus (HCV) now cures over 95% of such patients instead of the lifetime persistent infection their predecessors suffered. Additionally, the risk of developing liver cancer and needing a liver transplant is greatly diminished. Although anti-HCV therapy is initially expensive, compared to the former long-term hospitalization, likely transplant surgery, or cancer treatment, the long-term cost is modest and the value is great for patients as well as for the health care system.

In view of the successful outcomes with antiviral therapies, finding one or more molecules to combat ongoing Ebola infection seemed possible. An announcement had appeared that the small molecule, GS-5734,[1] an adenosine analog with antiviral activity, protected 100% of rhesus monkeys 3 days after initiation of a lethal Ebola infection. The virus tested was the Makona variant of Ebola, a virus isolated during the outbreak in Sierra Leone during 2014. Of further interest was the drug's effectiveness against other filoviruses (Marburg), arenaviruses (LASV), and coronaviruses (SARS). Despite its promise, though, this therapy still required testing for safety and, then, clinical trials in humans. The issues of large-scale manufacturing and pharmacokinetics still needed resolution to determine whether a drug, if produced, would be available for the next Ebola outbreak. It was not yet available for Dr. Khan.

The second approach in the development of treatments for human viral diseases is harnessing the host's immune response. The immune system has evolved to enable the host to resist invasion by organisms like, in this case, Ebola virus. Proteins in Ebola that trigger an immune response are called antigens (immunogens). A host's immune response to antigens can travel down two very different pathways. The most common and satisfactory one provides protection, controls the infection by either preventing it totally or lessening (attenuating) its effects, and induces a protective immune response. Such a response can provide long-term protection from Ebola virus so that a repeated infection does not cause disease—sort of one and done. To mimic that scenario, a vaccine is created to prime the immune response by programming it to recognize and then rapidly resist the Ebola infection in an individual who later becomes exposed to a person, animal, or substance contaminated with Ebola virus. Vaccines developed against viruses like measles, mumps, smallpox, yellow fever, poliomyelitis have changed the human and medical landscapes in that they succeeded in reducing the morbidity and mortality of human viral diseases. This success stands as one of the greatest of public health advancements. For example, smallpox alone in the 20th century killed an estimated 300 million individuals, about threefold more than all the wars of that century, including World Wars I and II. Vaccination eliminated smallpox so that, in the 21st century, not a single case has emerged.[2] Similarly, if one is infected by Ebola and survives, that person is immune to reinfection. Indeed, during the 2013−16 Ebola outbreak, such "naturally immune" individuals, because of their resistance to reinfection by Ebola, often worked to provide care and transport patients.

How does the immune response evade or control viral infection? The immune system must discriminate between foreign antigens, such as viral proteins, that are not found in humans (nonself) and those antigens that are self (proteins in your own cells and tissues). Cross-reactivity of a person's immune response to the virus with the individual's own "self-proteins" can lead to an autoimmune (antiself) response—a response to *self*-components, and autoimmune diseases like lupus, multiple sclerosis, diabetes, or thyroiditis.[3]

After an initial exposure to viral infection, the so-called acute phase, a race is on between the virus, which is replicating rapidly, and the host's immune system, which functions first to limit the amount of virus made and second to clear the virus from the host. At stake is whether the virus can successfully replicate itself. To combat the virus, the host mobilizes and uses many weapons, that is, both the immunologically specific and nonspecific responses. The nonspecific factors are all early combatants against the virus and the cells it infects. Included in this group are natural killer lymphoid cells, phagocytic macrophages—large cells that ingest or eat viruses—and proteins in the blood called complement factors that are capable of interacting with viruses and also destroying virus-infected cells. Important is the innate immune system that provides the initial defense against pathogens and primes the subsequent adoptive (T cell and antibody) immune response. The major players in the innate immune response are toll-like receptors, which recognize particular microbial patterns, and type 1 interferons (IFNs). Again, there are conflicting reports of whether IFNs are suppressed or exaggerated. Nevertheless, these innate systems are mutually complementary and are involved in developing the ensuing adoptive immune response. Type 1 IFNs upregulate molecules on cells that present major histocompatibility complex (MHC) molecules, molecules that code for self. MHC molecules are essential for optimal interaction with T cells. Following a virus infection or vaccination, major antigen-presenting cells (called dendritic cells) present segments of viral proteins (peptides) that become located within MHC molecules to naïve T cells, an action termed "priming." By this means virus-specific T cells are generated and expanded numerically. These T cells are made in the infected host for the specific control of a virus infection. Thus, the major combatants against viruses are antibodies and T lymphocytes, both of which mount the host immune response to Ebola, although the relative contribution of each is not unknown. However, in this battle, within 10−14 days after infection, either the replicating viruses or the host's immune response will emerge as the winner. If the immune response wins, viruses are vanquished, and the host survives with enduring immunity to that virus. However, if the immune response is overcome, the Ebola virus infection ends in the host's death or in a subset of individuals fingerprints of Ebola that can be found months later.[4,5]

What is the cellular (T cell) virus-specific immune response? The component parts are CD8+ and CD4+ T cells. The T stands for thymus-derived, and CD8+ or CD4+ indicates a specific molecule present on the cells' surface used to identify the cell. The thymus is a two-lobed gland of the lymphoid system located over the heart and under the breastbone. Lymphocytes formed in the bone marrow (hemopoietic stem cells) migrate to and enter the thymus where they are educated (mature) and are then selected to become either CD8+ or CD4+ T cells. CD8+ T cells function as surveillance and killer cells, which accounts for their name "cytotoxic T lymphocytes" (CTLs). They travel along the highways of blood vessels and wander among tissues throughout the body seeking cells that are foreign (not like self), because they express viral proteins. CTLs then recognize, attack, and kill such cells. By this strategy they eliminate the factories making viruses. CD8+ T cells also release soluble factors like interferon (IFN)-γ and tumor necrosis factor-alpha (TNF-α) that also have antiviral effects but do so without killing the virus-infected cell. CD4+ T cells predominantly serve a different role. They release soluble materials (proteins) that help or induce bone marrow-derived (non-thymic-educated) B lymphocytes to differentiate and make antibodies. CD4+ T cells release soluble factors (cytokines) that also participate in clearing a virus infection and, uncommonly, CD4+ T cells may also kill virally infected cells.

In contrast to reacting against cells, antibodies react primarily with viruses in the body fluids and are, therefore, most effective in limiting the spread of virus through the blood or in cerebrospinal fluids, fluids that bathe the brain and spinal cord. By this means, antibodies decrease a host's content of virus and diminish infectivity. Antibodies lower the number of viruses thereby allowing CTLs to work more efficiently. Antibodies along with effector molecules like complement can kill virus-infected cells. However, this mechanism is relatively inefficient compared to CTL killing of virus-infected cells. Although over 100,000 or more viral antigens must be present on the surface of a virus-infected cell to achieve lysis by antibody, less than 10 viral antigens expressed on a cell and restricted by MHC are all that are needed for a CD8 T cell to do the job. Thus, during infection, the eradication of virus-infected cells is the primary job of CTLs, whereas antibody's main task is to curtail the spread of virus in body fluids.

Antibodies are made by differentiated B lymphocytes named plasmacytes. Once activated, such cells can pump out 100 million antiviral antibody molecules per hour.

Antibodies latch onto and neutralize viruses by one of several mechanisms: (1) antibodies can coat or block the outer spike protein of the virus that is required to attach to the cell and begin its entry into the cell. By this means antibodies can prevent infection. This is the major action of vaccination. (2) Antibodies can aggregate or clump viruses so that the net number of infectious particles is reduced. (3) With the assistance of complement, antibodies can lyse (disintegrate) viruses, and (4) antibodies can react with viral antigens on the outer membrane of the infected cells to limit the manufacturing or transcription of virus molecules inside the cells, thereby restricting the amount of viruses made. Each antibody molecule generated acts on a specific antigen or target molecule of the virus.

When a host is initially exposed to an infecting virus or to a vaccine-containing viral antigens, antibodies specific for that virus as well as CTLs are generated. The CTL response is initiated on the first day of infection, expands over 100,000 to 1,000,000 times by doubling roughly every 12 hours with peak expansion 7−8 days after exposure. Thereafter, the quantity of these cells contracts and is maintained at 1%−2% of the total generated: these become immune memory cells. Immune memory T cells are rapidly stimulated and accelerated to respond when the same or cross-reactive infection occurs, i.e., they protect against a repeated infection by the same virus. Antibody responses peak after the CTL response, and unattached or free antibodies are often weakly detectable during the acute phase of infection. The number of antibodies then rises over a period of 2−4 weeks after infection, and they linger for years.

During an onset of acute viral infection, the mechanics of obtaining sufficient virus-specific CD8 T cells in humans to transfer into a MHC-matched individual is barely doable, even when attempted as a research study in a sophisticated clinical research center in an optimal hospital setting in the West. This task is not feasible during an Ebola infection in Africa. Yet, the transfer of convalescent plasma (plasma is blood from which red cells have been removed) harvested from Ebola-immune donors into acutely ill Ebola patients is possible. Such "immune" plasma (plasma containing antibodies to Ebola) can be placed in storage and thus remain available. The million dollar questions are—does this protocol work and, if so, is it of therapeutic value. For such transfers, blood is harvested in heparinized tubes to prevent clotting. T cells, red blood cells, macrophages, etc., are removed by centrifugation, and the resultant plasma is injected intravenously into the infected patient. In clinical trials with 84 patients of various ages acutely infected with Ebola virus, 200−250 mL of inoculum was tested. This plasma was harvested from previously infected Ebola patients who survived the disease.[6] Unfortunately, though, the survival rate of patients in the trial did not improve significantly over that of the controls who were not so-inoculated. However, the study was flawed in that neutralizing antibodies to Ebola were not quantitated or adjusted in the transferred plasma. Thus, exactly why the trial failed is not clear; was the entire procedure flawed or was the concentration of neutralizing antibodies in the plasma simply too low.[6] In other studies, potent neutralizing antibodies against Ebola were isolated from B cell clones of immune individuals who had recovered from Ebola infections. One such antibody, mAb114, given only once, protected 3 of 3 macaques even when administered as late as 5 days into the infection cycle.[7] The one macaque given no neutralizing antibody died at day 10 postinfection with 10^8 logs of virus. Similarly, potent neutralizing antibodies were isolated from a survivor of the 2014 Ebola outbreak, and 77% of the 349 monoclonal antibodies isolated neutralized Ebola indicating that a broad diversity of B cell clones target sites on the Ebola glycoproteins generated during infection.[8] Impetus to develop this strategy further came from these results with Ebola, and the same method was used in other animal/virus models. For example, monoclonal antibodies proved to protect guinea pigs from hemorrhagic arenavirus (Junin) infection of Argentina[9] and anti-HIV-1 antibodies prevented that infection in a monkey model.[10]

An important consideration underlying the proposed development of an Ebola vaccine was the fact that human survivors of the initial infection resisted reinfection when later exposed to Ebola. Such observations provided the scientific rationale and the logic for financial investment by government agencies like the National Institutes of Health (NIH) and charitable foundations to obtain an effective vaccine against Ebola. Less attention came from large pharmaceutical companies whose perspective was commercial. That is, even if successful, the vaccine would be used in as yet, nonindustrial (still third world) countries; therefore, the market for that vaccine would be smaller and less profitable than in industrialized Western countries. Also, because of its highly lethal nature, Ebola would require specialized facilities for handling and testing. Finally, insufficient data were available about the early events in human processing of Ebola virus infection.

As vaccines were considered, four Ebola-infected survivors undergoing the postacute phase of infection had been air-lifted from West Africa to Emory University School of Medicine. The profiles of these subjects' blood were analyzed at Emory and at the CDC in selective BSL-4 facilities. Surprisingly, the individuals displayed exceptionally strong anti-Ebola T and B cell responses.[11] These observations were the opposite of early reports that T and B responses were suppressed. The cause of this difference could be temporal, that is, the timing of blood sample collection, the advanced methodology used,[11] or perhaps the survivor population studied. Analysis of specific anti-Ebola T and B cell responses during the early acute phase of disease is currently unresolved. With new and better facilities in West Africa, especially Sierra Leone and Kenema Government Hospital (KGH), recent NIH support to identify and map T cell epitopes (regions on the virus as it comes to the surface of infected host cells) and funding for virologists and immunologists from both Sierra Leone and the United States, solutions should soon be forthcoming. The importance will be not only in identifying which arms of the host immune response function during Ebola infection but also in discovering the essential viral proteins to constitute a vaccine that provides optimal immunity and protection.

This background brings us back to the advancing death of Dr. Humarr Khan from Ebola virus infection, the controversy about using ZMapp for his treatment and the ethics of administering an untested drug to severely ill persons. ZMapp was a cocktail of monoclonal antibodies, the first to be suggested 2 years earlier as a potential transfer antibody therapy for Ebola.[12] ZMapp, created through international cooperation, showed promising results in primates but had not been tested on or approved for human usage. The majority of research to develop ZMapp was funded by the National Institutes of Health and Public Health Agency of Canada. When the Ebola outbreak began, the question of its use was raised, although it had not yet been deemed safe or effective for treating humans. Tests with monkeys infected with Ebola had shown a protective effect when they were given ZMapp at 3−5 days after infection.[12] Could ZMapp have a similarly protective effect in humans? Also, ZMapp had been produced in tobacco plants, a procedure used to expand its production. Would there be an issue of sensitivity to tobacco antigens for humans in therapy provided intravenously?

"The evidence presented here suggests that ZMapp offers the best option of the experimental therapies currently in development for treating EBO-V (Ebola virus) infected patients. We hope that initial safety tests in humans will be undertaken soon" Gary P. Kobinger of the Canadian Public Health Agency's National Laboratory for Zoonotic Diseases and Special Pathogens stated.

As Dr. Khan lay dying in the Doctors Without Borders Care Center in Kailahun, and the debate of whether or not to treat him with ZMapp was raging, over 480 miles to the Northwest in Monrovia, Liberia, at Samaritan's Purse ELWA Hospital, another crisis was unfolding. Samaritan's Purse, a Christian Relief Ministry, was coming to grips with the fact that two of its health care workers involved in treating Ebola patients, Kent Brantly, a 33-year-old physician, and Nancy Writebol, a care provider and missionary, showed clinical symptoms and signs of Ebola. The infection was confirmed by blood test. A hospital administrator for Samaritan's Purse, Lance Plyler, knew there were experimental drugs being developed elsewhere to combat Ebola. Losing no time he contacted a CDC official stationed in Monrovia to obtain names and places in the United States to seek the necessary therapeutics. Meanwhile, the infected and quarantined Kent Brantly was surfing the net with his laptop computer for a potential and possible therapy for Ebola. Brantly came across the report in *Nature*[12] that ZMapp, a cocktail of antibodies to neutralize Ebola, saved monkeys challenged with a lethal dose of Ebola even when they were several days into their illness. However, ZMapp had been tested only in cultured cells or animals and was not yet known to be harmless or effective in humans. Nevertheless, Brantly selected ZMapp as his choice. Was the material available? Lance Plyler contacted Gary Kobinger in Canada who was involved in the creation and testing of ZMapp. Plyler requested that the drug be sent to Samaritan's Purse Hospital for its immediate use. Kobinger informed Plyler that the nearest supply of the drug was only 480 miles away from the ELWA facility at the Doctors without Borders Kailahun Care Facility, where Khan was near death. Concerning Khan, the debate continued whether or not he should receive ZMapp. Plyler requested the drug and arranged for a helicopter to take ZMapp from the Doctors Without Borders Care Center to Monrovia.

The issue was the limited number of ZMapp treatments available. Only five doses existed in all of Africa and who should be selected to receive it. ZMapp had been sent to West Africa to the Doctors Without Borders Care Center in Kailahun primarily to determine how the drug would hold up in the African environment of heat and electrical power failures. The plan was to then return the drug to Canada and re-evaluate its antiviral efficacy: was it stable; was there a loss in potency? Although a promising anti-Ebola therapeutic, ZMapp had not attracted robust funds for its development. In a CBS News interview, Dr. Jeffrey D. Turner, President and CEO of Defyrus, a private life sciences biodefense company that collaborated with public government public health agencies and military partners in the United Kingdom and Canada, stated "The challenge that many people don't appreciate is that our plans were to scale up this drug for 2015 and even then, small amounts for clinical trials. What has really happened with this outbreak is it's caught us in a position where we didn't have enough ZMapp available because no one would have bought it."

ZMapp was produced in a specific type of biologically engineered, genetically modified tobacco plant, *Nicotiana benthamiana*, grown at Kentucky BioProcessing in Owensboro, Kentucky. The genes encoding the monoclonal antibodies made elsewhere were inserted in these plants and the plants then produced large quantities of the antibody. However, the procedure required several weeks. Thus, in this environment the manufacture of ZMapp took up to several months. Production was comparatively inexpensive, since as many as 100 million doses of antibody could be made for $36 million. In addition to Kentucky BioProcessing, the biopharmaceutical companies involved in the development of ZMapp included San Diego-based Mapp Biopharmaceutical and Texas-based Caliber Biotherapeutics.

After Humarr Khan's infection with Ebola was diagnosed at KGH, he asked for a transfer to the Doctors Without Borders Clinical Center in Kailahun to avoid further demoralizing his hospital staff. By chance, Kailahan was the site where ZMapp was stored. The question being debated by the doctors at that time was, should ZMapp be used for humans and, if so, in whom? How would these individuals be selected? It was into this setting that the severely ill Khan's name entered the debate. Initially and under pressure from the Sierra Leone government to do something for Khan, already recognized as a national hero in that country, the physicians treating Khan and others involved were presented with an ethical and philosophical dilemma. What if the patient died as a result of an allergic reaction to ZMapp? ZMapp had not undergone clinical trials for safety and efficacy in humans. What if the treatment failed? Many believed that Khan should receive ZMapp because, in all of West Africa, he was the leading and best known figure involved in the war on Ebola. However, others were hesitant. Also, the fact that Khan was African brought attention to the recent uproar that Westerners were subjecting Africans to lethal therapy.[13]

After a heated argument, the authorities in charge decided against giving ZMapp to Khan. That decision would be unpopular and not in line with the general requests for help advocated by Khan's government and specific requests of his colleagues. Garry and Sabeti applied pressure that he should receive the therapy. Khan's blood had already generated antibodies to the virus, indicating that his own immune system was beginning to work to combat the viral infection. However, as discussed above, early reports that the virus could suppress the immune system remained under reassessment, generating concern that the drug might affect his immune response.

Khan was denied the ZMapp therapy but not told of its potential benefit or that it was on hand. Representatives of Doctors Without Borders said he wasn't consulted because it would be unethical to inform him of the potential drug that was not available to him. One of Khan's close friends and fellow clinical researcher, Dr. Daniel Bausch,[14] strongly disagreed, "Dr. Khan is the ideal person to make an informed decision, and I feel strongly that he should have been asked if he wanted it or not...that's one area where, frankly, I am critical." Nevertheless, shortly after being denied the drug, Khan died.

The next day, despite the same concerns that prevented ZMapp treatment for the dying Khan, other infected individuals received ZMapp. The treatment that might have saved Khan's life was, instead, transported to Samaritan's Purse ELWA Hospital in Liberia and administered to two Ebola-infected health workers from the United States. This act created a strongly emotional response from the local population and the international community. Giving ZMapp to two Westerners who survived and not to Khan, an African who died, was highly controversial. Since not all those infected with Ebola die (mortality ranging 50%–70%), possibly the two Westerners would have survived without ZMapp. After all, many did. If Khan had been given ZMapp and died, would Africans be convinced he was used as a guinea pig to be tested with an experimental/unproved therapy without either formal local approval or approval by Western scientific and government committees? The two American recipients of ZMapp, Kent Brantly and Nancy Writebol, were members of Samaritan's Purse, the Christian Missionary in Liberia. A source from the NIH stated that someone from the CDC contacted Samaritan's Purse and that an NIH scientist later informed them of the drug. However, no one knew if the drug would work or if the ultimate recovery of those taking the ZMapp was due to the drug. After receiving ZMapp, Kent Brantly showed remarkable clinical improvement, although, in contrast, after receiving the drug Nancy Writebol's condition worsened. However, both Brantly and Writebol survived, but may have recovered without the ZMapp—one does not know. The indisputable fact is that Humarr Khan died without receiving ZMapp when, with its therapeutic effect, he might have lived.

Many of the natives in West African lacked trust in international efforts to combat disease and Ebola. Strangely, since the reported outbreaks of mass Ebola virus infections, some groups of native Africans have expressed doubt that Ebola really exists. Others have voiced allegations that health officials purposely infected the populations to harvest their organs. Hospitals, health care stations, and workers have been attacked and stoned by mobs. Awareness, quarantine, and prohibition of touching people sick or dead from the disease have been viewed by many Africans as myths of Western propaganda. The African custom of touching and washing the body of a dead person prior to burial was being denied and led to resentment, demonstrations, and riots. Further, such public health measures are alien to their culture. Many West Africans believe in cultural superstitions and view Ebola as a curse rather than a pathogen, associating Ebola with witchcraft and sorcery brought there by foreigners. Some locals believe that doctors are killing Ebola patients as a punishment for sexual promiscuity. Fabio Friscia, UNICEF coordinator of the Ebola awareness campaign, explained that what was creating the greatest problem in controlling Ebola was the "behavior of the population."[15]

In Guinea, Liberia, and Sierra Leone the local population often attacked and disrupted health care workers, forcing them to leave treatment centers and hospitals. In one episode occurring in South-East Guinea, eight members of an Ebola disinfection and awareness team were killed with stones and machetes by fearful villagers. The members killed were part of a relief wing of the Christian and Missionary Alliance.

In August 2014, armed locals attacked a medical clinic in Liberia, in an area named West Point, where patients were quarantined. Locals broke down doors, stole bloody mattresses, sheets, and equipment and caused patients to flee in a panic. The day before the raid, a crowd of several hundred locals drove away burial teams and police, chanting "No Ebola in West Point."

Violent attacks have resulted in Doctors Without Borders and medical volunteers having to withdraw from their posts because of concerns for their safety. A spokesperson for Doctors Without Borders stated "We understand very well that people are afraid because it is a new disease here, but these are not favorable working conditions so we are suspending our activities." [16]

Despite such difficult times, the story of ZMapp is not over. As both foreign as well as African health care providers and doctors continued to die from Ebola virus infection, the WHO endorsed the use of ZMapp to combat the uncontrolled outbreak. Liberian President, Ellen Johnson Sirleaf, in a direct request to President Obama, asked for a supply of the drug to be used for the treatment of local doctors.

Representing the producers of ZMapp, a spokesperson for Kentucky BioProcessing stated, "Though this is all relatively new and there's still a long way to go and a lot of things are going to happen as we go into drug-approval protocols with ZMapp...I think certainly it shows great promise that the tobacco plant can be utilized for such things. We'll see where that goes. We're certainly optimistic."[17]

REFERENCES

1. Warren TK, et al. *Nature* 2016;**531**:381−3.

2. Oldstone MBA. *Viruses, plagues, & history*. New York, NY: Oxford University Press; 2010.

3. Oldstone MBA. *Molecular mimicry: infection inducing autoimmune disease. Current topics in microbiology and immunology*. Heidelberg: Springer-Verlag; 2005.

4. Heeney JL. Ebola: hidden reservoirs. *Nature* 2015;**527**:453−5.

5. Chughtau AA, Barnes M, Macintyre CR. Persistence of Ebola virus in various body fluids during convalescence: evidence and implications for disease transmission and control. *Epidemiol Infect* 2016;**144**:1652−60.

6. van Griensven J, et al. *N Engl J Med* 2016;**374**:33−42.

7. Corti D, et al. *Science* 2016;**351**:1339−42.

8. Bornholdt ZA, et al. *Science* 2016;**351**:1078−83.

9. Zeitlin L, et al. *Proc Natl Acad Sci* 2016;**113**:4458−63.

10. Gautam R, et al. *Nature* 2016;**533**:105−9.

11. McElroy AK, et al. Human Ebola virus infection results in substantial immune activation. *Proc Natl Acad Sci USA* 2015;**112**:4719–24.

12. Qu X, et al. *Nature* 2014;**514**:47–53.

13. Omonzejele PF. Ethical challenges posed by the Ebola virus epidemic in West Africa. *J Bioeth Inq* 2014;**11**:417–20.

14. *Dr. Daniel Bausch knows the ebola virus all too well.* NPR, KPBS; September 22, 2014.

15. Basu, Indrandi. *The superstition that is dramatically escalating the ebola outbreak* <http://the-week.com/articles/444857/superstition-that-dramatically-escalating-ebola-outbreak>; August 4, 2014 [accessed 27.07.16].

16. FoxNews.com. *Fox news* <http://www.foxnews.com/world/2014/04/05/crowd-attacks-ebola-treatment-center-in-guinea.html>; April 5, 2014 [accessed 27.07.16].

17. Briggs, Bill. *Ebola treatment: how big tobacco and the military came together—NBC news.* <http://www.nbcnews.com/storyline/ebola-virus-outbreak/ebola-treatment-how-big-tobacco-military-came-together-n173311> [accessed 27.07.16].

CHAPTER 6

Robert Garry: Managing the Effort to Curtail Ebola's Curse

On May 23, 2014, the first person suspected of carrying Ebola arrived at KGH. The patient, a woman, had suffered a miscarriage and was clinically sick with bleeding and a high fever while complaining of a headache and backache. She had recently attended the funeral of a traditional healer, who was well-known in the area for her treatments and medicinal skills that cured local diseases.[1] Shortly before her death, the healer had been ministering to individuals from Guinea who suffered from the symptoms of hemorrhagic fever. Soon afterward, she became ill herself and died. Although that death was not yet attributable to Ebola infection, physicians at KGH knew of the Ebola outbreak raging in Guinea. Still, no-one they knew of had yet been affected in Sierra Leone. Nevertheless, precautions were taken at KGH in case individuals infected with Ebola were to cross the shared border of Guinea/Sierra Leone. The previous March, Kristian Andersen of Harvard/MIT's Broad Institute had set up testing with polymerase chain reaction (PCR), a method of identifying Ebola DNA or RNA in blood. Anderson anticipated that Ebola infections occurring in Guinea would eventually appear in Sierra Leone, and this test would be ready to detect Ebola RNA. The wait was not long. Within just a few days after the diagnostic test for Ebola was designed and implemented at KGH, the "mysterious deadly infectious disease" occurring in Guinea was definitively shown to be Ebola.[2] That result was posted on the internet March 23, 2014, and the test was available at KGH within a few days. First, a blood sample sent down from the village of Koindu tested positive in the KGH laboratory by Augustine Goba. Then, 2 months later, the woman who had had a miscarriage, and first suspected infected patient, arrived at KGH. A sample of her blood was tested by Goba and was positive for the Ebola virus infection.[3] A portion of the blood sample sent to the Broad Institute was subsequently confirmed by Pardis Sabeti's group there as containing Ebola

Ebola's Curse. DOI: http://dx.doi.org/10.1016/B978-0-12-813888-5.00006-8

virus. The patient was promptly placed in the KGH Lassa isolation ward. Information about this first confirmed Ebola patient reached Bob Garry in the United States. Garry was the person largely responsible for set-up, maintenance, and training at the LASV ward while working with Humarr Khan. Garry was also in charge of maintaining the diagnostic and research infrastructure at KGH and in the surrounding Sierra Leone area. He rapidly left New Orleans, arriving at KGH to ensure that the staff was prepared, properly outfitted with protective garments, and using masks. He began collecting blood to be used for diagnosis of patients as well as preparation of samples to be sent to Sabeti's group for virus genome analysis. This procedure was accomplished at the start and throughout the Ebola rampage in the endangered African areas.

The blood of infected victims contains high titers of Ebola virus, and contamination by this highly infectious agent can cause the specimen's handler to suffer a lethal outcome. After its collection from patients, such blood is treated with chemicals to inactivate the Ebola virus' infectivity but preserves sufficient viral RNA for sequencing and serum proteins for analysis. In addition, these blood samples are studied to characterize the infected individual's immune response to the infection as well as to perform the chemical assays necessary to treat the patient and track his/her infection for clinical care. Samples sent out of Sierra Leone for analysis had to, first, pass a critical human subjects' review and then receive permission to exit the country by the KGH hospital committee and the Sierra Leone government. Bringing samples into the United States for analysis also required a comprehensive review by the CDC, NIH, and the institute where the samples would be sent before performing their specific assay. In this case, official approvals had been awarded to the Tulane Medical Center (the location of Garry's team for antibody analysis) and to the Harvard and MIT Broad Institute (Sabeti's team for viral genome analysis). Currently, The Scripps Research Institute in La Jolla, California (Oldstone, Sullivan, and de la Torre teams) has also received such approval from their institute's human subjects research review committee as well as from the NIH to study innate and adoptive immune responses of individuals infected by Ebola. Obtaining permission to ship and study samples has been and is labor- and time-intensive and requires persistent but competent workers who can tell a truthful and accurate story with appropriate details.[4] In terms of overall organization, Bob Garry was the right person at the right time to accomplish these details working with both African and American officials and representatives.

Bob Garry was born in Terre Haute, Indiana, obtained a Bachelor of Science degree in Life Sciences from Indiana State University and a PhD in Microbiology from the University of Texas at Austin. After postdoctoral training and a junior faculty appointment in the Department of Microbiology, Garry was recruited by Tulane University School of Medicine at New Orleans, Louisiana, where he rose to the positions of Professor of Microbiology and Immunology and Director of the Interdisciplinary Program in Molecular and Cellular Biology.[5]

A major component of Garry's responsibility was managing a consortium of scientists who were and are developing countermeasures against African hemorrhagic viruses, including diagnostics, immunotherapeutics, and vaccines. One of Garry's defining accomplishments was the development of immunoassays with high sensitivity and specificity for Ebola virus.[6] The Ebola rapid diagnostic test devised by Garry and colleagues is the only immunoassay to receive emergency use authorizations from both the FDA and WHO. "It's a rapid test that is able to indicate if the Ebola virus is present in 15 minutes or less. It's something we have been working on for several years with Lassa fever, so it was much easier to put together a program where we can develop the test which works very well," explained Garry. However, bureaucratic difficulties arose thus taking longer than necessary for the test to be used. "This test can really make a difference in containment by placing people in the right holding facilities. The WHO is now running a trial and it is working just the way we had hoped. It's a rapid test that is able to indicate if the patient is or isn't affected."

The Eastern Province of Sierra Leone has the world's highest annual incidence of Lassa fever[7] and the largest number of persons infected with Ebola during the 2013–15 outbreak.[8] KGH was an important site for clinical and laboratory research on Lassa fever throughout the 1970s and 1980s. The violent civil conflict there, sometimes referred to as the Blood Diamonds War, broke out in 1991 and did not end until 2003.[9] This clash forced suspension of Lassa fever research that was being done at the CDC center established at the Nixon Memorial Hospital at Segbwema, which is roughly 62 miles from Kenema. This CDC field station and laboratory had been established by Joseph McCormick.[10] Because that site was unsafe for medical personnel during the civil war, part of the CDC's effort was moved to the KGH center in Sierra Leone. Later, following the cessation of hostilities, a consortium of Lassa fever researchers led by Garry began rebuilding the scientific infrastructure at KGH. Then, in response to the Ebola outbreak and in anticipation of a future epidemic, work was being completed on a new viral hemorrhagic fever (VHF) ward of 40 beds, an increase over the original 14 beds, and with improved isolation units and patient waste disposal systems. The finished complex will be named the Khan Center of Excellence at KGH. This name celebrates the late Dr. Sheik Humarr Khan who was the director of the KGH Lassa ward and died there after contracting Ebola from his patients[11] (as described in Chapter 4: Sheik Humarr Khan: Leading the Fight Against Ebola in Sierra Leone at Kenema Government Hospital). Because of its historic contributions to VHF research and the prominent role it played in the West African Ebola outbreak, KGH has been established as a Center of Excellence for work on VHFs. Presently, Bob Garry is Director of the International Center of Infectious Diseases Research. Through support of NIH's Human Health and Heredity Program (H3Africa) as well as the Wellcome Trust and World Bank, the KGH center is expanding training opportunities for West African scientists in infectious diseases and genomics research. Garry was also a member of the leadership teams that founded the Viral Hemorrhagic Fever Consortium (VHFC),[12] and the African Center of Excellence for Genomics of Infectious Diseases (ACEGID),[13] which is part of the H3Africa Consortium.[14]

The West African Ebola outbreak provided the opportunity to test the facilities and assays developed by Garry, Andersen, Sabeti, and their colleagues. The Ebola outbreak began in December 2013, originating in a region of forested Guinea located only about a 3-hour drive from KGH. The identity of this fever was established by genome sequencing.[3,15] Sick individuals began arriving at the KGH VHF clinic 2 months later, and there, for the first time, an advanced clinical and laboratory infrastructure was available onsite for thorough study of viral hemorrhagic fevers. Also at KGH, diagnosis of the first case of Ebola in Sierra Leone emerged. In addition, VHFC scientists led by Pardis Sabeti at the Broad Institute now received samples for the analysis of Ebola virus genetics/genomics as the virus spread into and through Sierra Leone.[3,15] That study demonstrated a single introduction of Ebola virus into a human recipient with subsequent human-to-human spread,[3,15] and documented a rapid accumulation of mutations in the viral genome. VHFC and others extended those studies by analyzing Ebola virus genetic variability as it spread through the populations of West Africa.[15,16] VHFC also performed the largest and most in-depth clinical study and provided the best written description of Ebola virus disease to date.[17] The clinical studies showed that, in addition to the expected findings of fever, lethargy, and headache, this West African variant of Ebola virus caused a gastrointestinal illness that focused attention on a newfound clinical aspect of the disease. Bleeding manifestations that predominated in prior Ebola outbreaks in central Africa[17] occurred but with a lower frequency than before ($\sim 1\%$ of patients). Basic studies of the gastrointestinal malfunction revealed that the Ebola virus delta peptide functioned as a viroporin or an enterotoxin with a likely role in Ebola virus pathogenesis. Thus, for the 2013–15 Ebola outbreak, Garry's work with the Sabeti team was essential in determining where and when the viral spread took off. Genomic analysis of the Ebola viruses isolated throughout the outbreak documented that the incidences of Ebola during 2014 in Sierra Leone originated from the Ebola virus isolated earlier in Guinea.[3,15,16,18] Spread to humans in Sierra Leone came from individuals who attended the traditional healer's funeral, and the infection they carried migrated to Sierra Leone. From there, the virus spread uncontrollably throughout Sierra Leone. This conclusion derived from sequences of 99 viral genomes from 78 Ebola-infected individuals.[3] Garry commented..."This is the first study to document deep viral genomics during a human outbreak of a hemorrhagic fever like Ebola." "We get a close look at not only how the virus is evolving as it passes from one person to the next, but also how the virus changes as it replicates within a person."

With confirmation of this first case of Ebola virus infection in Sierra Leone at KGH, Garry returned there and brought 28 cases of personal protective equipment including full suits and face masks (Fig. 4.1). His instructions insisted that all health care providers who see or treat Ebola-infected patients must wear these items. Because the virus is so highly contagious ..."we tell them to wear gloves and to protect their eyes and we've shown people how to do a traditional burial, only wearing gloves. And you can allow the body to be washed briefly. Workers have been attentive to the traditions, allowing the body to be wrapped without exposing people to the virus"...Garry stated.

Khan and Garry had worked together for nearly 10 years and, over that time, developed both a professional and personal relationship. Khan was more than a colleague to Garry, he was a close friend. When speaking of his personality and work, Garry described Khan as "a national hero" and "tremendous human being." However, as Ebola's curse exponentially increased its spread in tandem with exposure to an over-worked and exhausted Khan, Garry pressed himself harder to find possible remedies. Simultaneously, he strove to raise awareness throughout the international community and to Americans as well as raising money, soliciting contributions for supplies, funding equipment, and attracting participation from physicians and nurses. Endlessly, he repeated, "We're working on vaccines and medicines for Ebola and other hemorrhagic fevers, the solutions are coming!"

As the outbreak spiraled out of control, Garry repeatedly traveled back and forth between the United States and KGH. "It's been an eventful six weeks. I would be going back (to KGH) except there are things that are needed that I can't do over there like be in communication with people who are funding the work and trying to get more funding and things like that."[6] But in the end, the epidemic subsumed over 11,000 individuals in Sierra Leone. The epidemic was reported as dying out in 2015. However, several new cases, from 20 to 50, occurred per month in 2016 and continues as Ebola simmers along. The WHO Director General reported an end of the epidemic on March 30, 2016, since no new cases were reported for 42 days after the most recent cluster on January 14, 2016. However, Ebola had not finished its work. Sporadically, new cases subsequently occurred. Dr. Khan and many of his original health care staff would not be up front to see and care for these newly Ebola-infected individuals. Khan and over 40% of his health care providers died from the first round of the battle against Ebola in KGH. Let's hope that the U.S. government and other governments and organizations have not abandoned the battle against Ebola, and they will not again engage too little or too late. That concern arises justifiably as $500 million dollars formerly appropriated for Ebola by the United States were recently withdrawn and sent to respond, instead, to Zika infection, despite Ebola's continued hold in West Africa.

Robert Garry had a deep personal and professional relationship with Khan. Their work together had substantial effects on the Ebola outbreak. With Khan, Garry managed the set-up, maintenance, and training of staff procedures for work on hemorrhagic fevers at KGH and traveled back and forth to Sierra Leone and the United States bringing necessary medication, gear, and equipment. His work on Ebola virus countermeasures, including diagnostics, immunotherapeutics, and vaccines, resulted in the success of a diagnostic test that swiftly detected Ebola virus in the blood of patients. He was accompanied among his journey before, throughout, and after the outbreak with Pardis Sabeti who became a close friend and colleague.

REFERENCES

1. Mueller, Katherine. *Turning to traditional healers to help stop the ebola outbreak in Sierra Leone.* IFRC. <http://www.ifrc.org/en/news-and-media/news-stories/africa/sierra-leone/turning-to-traditional-healers-to-help-stop-the-ebola-outbreak-in-sierra-leone-66529/>; July 31, 2014 [accessed 12.03.16].

2. Baize S, et al. Emergence of Zaire Ebola virus disease in Guinea. *N Engl J Med* 2014;**371**:1418−25.

3. Gire SK, et al. Genomic surveillance elucidates Ebola virus origin and transmission during the 2014 outbreak. *Science* 2014;**345**:1369−72.

4. *NIAID role in Ebola and Marburg Research.* NIH <http://www.niaid.nih.gov/topics/ebolaMarburg/research/Pages/default.aspx>; February 26, 2016 [accessed 14.04.2016].

5. *The people of Tulane Cancer Center Research.* Tulane University − School of Medicine − Robert F. Garry. <http://tulane.edu/som/cancer/research/people/robert-f-garry.cfm> [accessed 16.04.2016].

6. Catalanello, Rebecca. *Tulane researchers race to develop rapid Ebola finger-prick test.* NOLA.com<http://www.nola.com/health/index.ssf/2014/10/tulane_researchers_race_to_dev.html>; October 13, 2014 [accessed 3.4.2016].

7. Kenema Government Hospital. *Viral hemorrhagic fever consortium* <http://vhfc.org/consortium/partners/kgh> [accessed 20.5.2016].

8. *2014 Ebola outbreak in West Africa—outbreak distribution map.* Centers for Disease Control and Prevention <http://www.cdc.gov/vhf/ebola/outbreaks/2014-west-africa/distribution-map.html>; March 17, 2016 [accessed 14.06.2016].

9. *Conflict diamonds.* Amnesty International USA. <http://www.amnestyusa.org/our-work/issues/business-and-human-rights/oil-gas-and-mining-industries/conflict-diamonds> [accessed 1.06.2016].

10. Joseph McCormick, MD. University of Michigan School of Public Health <https://practice.sph.umich.edu/practice/dynamic/site.php?module = courses_one_instructor&id = 140>; 2003 [accessed 5.04.2016].

11. *In memory of Dr. Sheik Humarr Khan.* Africa Centre Of Excellence for Genomics of Infectious Diseases <http://acegid.org/index.php?active = page>; August 7, 2014 [accessed 20.04.2016].

12. *Viral hemorrhagic fever consortium* <http://www.vhfc.org/> [accessed 20.06.2016].

13. *Africa Centre of Excellence for genomics of infectious diseases.* <http://acegid.org/>; 2016 [accessed 20.06.2016].

14. Consortium. *H3Africa human heredity & health in Africa* <http://www.h3africa.org/consortium> [accessed 20.06.2016].

15. Goba A, et al. An outbreak of Ebola virus disease in the Lassa fever zone. *J Infect Dis* 2016;**214**:S110−21.

16. Tong Yi-Gang, Shi Wei-Feng, Liu Di, Liang Long, Bo Xiao-Chen, Liu Jun, et al. Genetic diversity and evolutionary dynamics of Ebola virus in Sierra Leone. *Nature* 2015;**524**:93−6. [accessed 14.06.2016]. http://dx.doi.org/10.1038/nature14490.

17. Schieffelin JS, et al. Clinical illness and outcomes in patients with Ebola in Sierra Leone. *N Engl J Med* 2014;**371**:2092−100. [accessed June 15, 2016] http://dx.doi.org/10.1056/NEJMoa1411680.

18. Sanchez A, Geisbert TW, Fieldmann H. Filoviridae: Marburg and Ebola viruses. In: *Fields virology*, 4th ed. Philadelphia, PA: Lippincott Williams & Wilkins; 2001. p. 1297−9.

FURTHER READING

Press release − VHFC researchers publish key findings on the 2014 Ebola outbreak. Viral Hemorrhagic Fever Consortium <http://www.vhfc.org/media/news/press-release-vhfc-researchers-publish-key-findings-2014-ebola-outbreak>; August 28, 2014 [accessed 20.06.2016].

CHAPTER 7

Pardis Sabeti: Geneticist Tracking Ebola's Travels and Changing Profile

At Kenema Government Hospital (KGH), a group that included Pardis Sabeti along with Humarr Khan and Bob Garry, all in leadership roles, gathered to establish an operational Viral Hemorrhagic Fever Consortium. This group of scientists and medical workers also represented Tulane University, Harvard/MIT plus others from West Africa and the United States, a staff dedicated to studying and combating arenaviruses (LASV).[1]

Pardis Sabeti, after her undergraduate training at the Massachusetts Institute of Technology, attended the University of Oxford where she obtained (in 2002) a PhD in Biological Anthropology. Her work was on "the effects of natural selection and recombination on genetic diversity in humans." Her studies examined genetic diversity in Africans and host susceptibility to malaria.[2] For that task, Pardis developed new algorithms to study natural selection.[3] Along with this preparation, which aided her eventual investigations on Ebola in West Africa, she first attended and received a medical degree from Harvard in 2006. After graduating from medical school, Pardis took postdoctoral work that continued her prior studies under the mentorship of the well-known geneticist and molecular biologist Eric Lander, head of the MIT Broad Institute in Cambridge, Massachusetts. During that time, Pardis continued research focused on the development of methodologies and software to detect natural selection in genome-wide studies and in scanning the human genome for evidence of natural selection in infectious diseases. Her academic accomplishments led to her recruitment for a faculty appointment at the Center for Systems Biology, Department of Organismic and Evolutionary Biology at the Broad Institute of MIT and Harvard in 2008. There, Pardis set up her own laboratory and worked to develop new analytic methods and use rapidly emerging genomic knowledge and resources to study evolutionary adaptation in humans and pathogens. The idea was to investigate and

Ebola's Curse. DOI: http://dx.doi.org/10.1016/B978-0-12-813888-5.00007-X

characterize functional changes over time that shaped humans' responses to selected pathogens. That focus led her to investigate the genetic diversity of LASV and how mutations in a host's cellular receptors might account for susceptibility or resistance to this virus. A cell surface receptor, in this case alpha-dystroglycan (α-DG) modified by the glycosyltransferase (enzyme) LARGE, leads to the attachment of LASV and its entrance into a cell, where it can then replicate its progeny.[4,5] From her observations of mutations in LARGE occurring in geographic areas of LASV infection, Pardis was attracted to countries in West Africa (Sierra Leone and Nigeria) where Lassa is endemic and where she joined work with the Viral Hemorrhagic Fever Consortium.[1] Pardis joined the scientists at KGH in 2008 both to help in the rapid diagnosis of individual patients infected with LASV as well as to uncover other pathogens. The idea was to provide warnings of local outbreaks before these might become a global threat. Thus, the team of Humarr Khan, Bob Garry, and Pardis Sabeti was formed, joined with other colleagues, and traveled to Nigeria to inaugurate the African Center of Excellence for Genomics of Infectious Diseases. The Viral Hemorrhagic Fever Consortium was instrumental in setting up the African Center of Excellence for Genomics of Infectious Diseases in Ogun State, Nigeria, an area where LASV infections were common. The Center's mission was to monitor dangerous infectious agents in West Africa.[6]

The unexpected outbreak of Ebola starting in 2013 and peaking in 2014—15 in Sierra Leone and other parts of West Africa turned her attention and commitment to investigating the diversity in Ebola viruses and examining how they evolved into an infectious agent that caused profound human morbidity, misery, and mortality. To quote the father of microbiology, Louis Pasteur ... "Chance favors the prepared mind."

A vivacious, intelligent, and determined scientific explorer, Pardis' American adventure began when she and her family fled the Iranian revolution in 1979, emigrating from Iran at age 2. In a whirlwind of academic achievements since then, in less than her current 40 years of life, she achieved a Rhodes Scholarship (1997—2000), Burroughs Wellcome Fund Career Award in the Biomedical Sciences (2006—13), and a recent Howard Hughes Medical Institute Investigator Award (2015-on). She received the Ellis Island Medal of Honor (2013),

Carnegie Corporation of New York Great Immigrants Award (2014), the Smithsonian magazine's American Ingenuity Award in the Natural Sciences (2012), and National Geographic Emerging Explorer Award (2013). For her work on Ebola virus, she was named one of TIME Magazine's Persons of the Year in 2014 (Ebola Fighters) and one of TIME Magazine's 100 most influential people in 2015. In addition to her life in science and teaching, Pardis is an active musician. Remarkably, she was and continues as the lead singer and bass player of the rock band "Thousand Days" that she helped form. The band received an Honorable Mention in the Billboard World Song Contest. The band and Pardis, besides performing and composing albums, also made/make music videos to spark and enlighten young people's interest and knowledge in science.[7]

Her work to determine whether a specific genetic variation in a given host gene is widespread in a population as a result of favored natural selection against a pathogen involves analysis and understanding factors that influence the response of humans to the pathogen under scrutiny as well as the genome of the pathogen. This is challenging work and in part reflects the influence of her father, Parviz Sabeti, a high ranking official in the late Shah of Iran's government and known for his expertise in taking difficult, meaningful jobs. According to Pardis . . ."(he) took one of the toughest jobs in the Iranian government because he cared about his nation more than himself. . .his courage and conviction have always driven me to want to make a difference – making the world a better place."[8] Thus, Pardis led a team to sequence the viral genome (Ebola and Lassa) from infected individuals at the beginning and during the infectious outbreaks. She was one of the first to use in-depth real-time DNA sequencing in the midst of a deadly epidemic.[9]

Obtaining 99 samples from 78 different patients in the first 3 weeks of the Ebola outbreak in Sierra Leone at KGH,[9] Sabeti and her colleagues proved genetic similarity between viruses sequenced at Sierra Leone compared to sequences from Ebola viruses isolated at Guinea where, in 2013, the outbreak began. The results suggested that the epidemic started with a single transmission into the human population from a natural animal reservoir and then continued by human-to-human transmission.[10] The human-to-human transmission both continued and extended the outbreak. Her results reinforced the mandate to medical and public health personnel that human contact with infected patients or animals must cease while providing the rationale to do so.

Later, sequencing data obtained by analysis of the Ebola genome in 232 blood samples taken over a 7-month period in Sierra Leone led Sabeti and colleagues to confirm that one strain of Ebola virus entered the country and continued to spread there. Further, the earlier genetic versions of Ebola genome in Guinea did not fade away, that is, they were not replaced by selective mutations. Instead, those strains continued to spread within Guinea to its capital city, Conakry, population over 1,668,000 people. Ebola virus isolated and sequenced in Conakry continued to be a close version of the Ebola first isolated in Guinea and Sierra Leone. Thus, the early lineage designated "an Ebola virus" was persisting, again pointing to the fact that all these viral samples descended from one another. All told, the accumulated data strongly indicated that the entire outbreak was likely caused by the introduction of a single virus from an animal reservoir.

Despite border closings among neighboring West African countries monitored by local police, soldiers, and the WHO, sick people from the Taï Forest area continued to cross back and forth into and out of Guinea to Sierra Leone and to Liberia.[11] The Taï Forest borders the west coasts of Guinea, Sierra Leone, and Liberia. The continuous border crossings and migrations of people yielded a second wave of infected individuals, in numbers exceeding those in the first wave. Guinea health officials and the WHO knew sick people had crossed into Sierra Leone in March 2014, long after the first case emerged in Guinea during December 2013. However, Ebola virus infection and disease were not detected in Sierra Leone until May 2014. The first laboratory evidence of Ebola in Sierra Leone appeared in May 2014 at KGH, where a patient's blood sample was tested for viral RNA. For this test, parts of RNA from the Ebola virus were used with RNA isolated from the patient's blood, which proved to contain RNA identical to that of the Ebola virus. This assay, called a polymerase chain reaction (PCR), was set up in March in anticipation that Ebola raging in Guinea could soon appear in Sierra Leone. When this prediction came true, the positive diagnosis led Sabeti to form a team with advanced diagnostic skills and equipment and send them to KGH. They were to assist in rapid analysis of the genomic sequence of Ebola and track its spread and evolution. Pardis commented, "The faster you can get a diagnosis of Ebola, the faster you can stop it...the big question is, is this going to be stopped?"

Attending a single funeral where numerous people were exposed to Ebola was the start of the Ebola epidemic in Sierra Leone. By simply touching the body of the dead local shaman, multiple individuals became infected. "Virus like a tidal wave was coming into the country [Sierra Leone]..."the first case was manageable" lamented Pardis.[12] Virus infections were expanding exponentially with a doubling period of 35 days. A major concern was to prevent Ebola from migrating to larger cities where large susceptible populations lived: Conakry in Guinea (over 1,667,000 population), Freetown in Sierra Leone (nearly 1,000,000), Monrovia in Liberia (over 1,000,000), and Lagos in Nigeria (over 17,000,000). Early questions arose from those involved in fighting to combat and stop the spread of the infection. First, had new Ebola virus variants emerged that would increase the human host range and, if so, would they cause more severe or less severe disease? Second, was the human immune response to fight the virus being compromised? The immune response might be compromised by any of several mechanisms. For example, Ebola might suppress the ability of the host's immune response to recognize and combat the virus. Ebola might, by mutation, change its own amino acids to avoid recognition by the host's immune response, i.e., Ebola variants that escape either virus-specific T cell recognition or virus-specific antibody recognition, both pillars of the adoptive immune response. The virus might influence the production of interferon (type 1), an early innate immune response made by the infected host to limit (interfere) with virus spread and protect uninfected cells from becoming infected. Production of interferon is known to be suppressed by one of Ebola virus' proteins, VP35. Alternatively, might Ebola virus cause an exaggerated or excessive immune response to enhance anti-Ebola T cell numbers and responses leading to an event known as "cytokine storm." The end result of a cytokine storm is attracting the body's scavenger and killer cells, thereby causing increased injury of tissues and killing of both Ebola-infected cells and even uninfected host cells. This is called immunopathology. In other viral infections, virus-induced immunopathology is known to enhance the clinical symptoms (what the patient feels and states) and signs (what the doctor finds and describes) due to tissue and cell injury as a consequence of an exaggerated antiviral immune response. Additionally, a cytokine storm can release large amounts of small protein molecules called cytokines or chemokines known to produce chemical reactions that attract macrophages, polymorphonuclear cells, and lymphocytes to a site. When these programmed host cells interact with infected and uninfected cells, morbidity and mortality can result from the virus infection. Cytokine storms are also suspected of playing a role during acute Ebola infection as shown in animal models of Ebola and by high levels of cytokines/chemokines in infected individuals. In addition, anti-Ebola viral antibodies alone or along with cells that bear Fc receptors or complement proteins can also cause tissue and cell injury.

With all eyes on the West Africa outbreak raging in 2014, Ebola attacked Central Africa again. In July 2014, Ebola reappeared in Central Africa where it was first recorded, the latest of 25 recorded outbreaks. This time, the Democratic Republic of the Congo (DRC) was the target, specifically, the rural village of Inkanamongo, which is located in the vicinity of Boende. Of 69 infected individuals, 57 died, an exceedingly high mortality rate of 83%. Among the dead were eight health care workers. Inkanamongo is located in a humid, tropical forest delineated by two large rivers and with poor entry and exit roads. This was the seventh Ebola outbreak in DRC. By contrast, over 28,000 cases of Ebola-Zaire occurred in the West African outbreak. Alarmingly, the Ebola-Zaire virus isolated from this 2014 outbreak in DRC was 99% identical to other Ebola-Zaire viruses isolated in the DRC years earlier, and >97% identical to Ebola viruses isolated in 2014 and 2015 from Sierra Leone.[13] How an almost identical Ebola-Zaire viral strain from the Congo in Central Africa could also be found in West Africa is not at all clear, since the two areas are separated by over 2400 miles. Independent earlier studies by D.W. Thomas of the University of Aberdeen and Heidi Richter of the University of Florida used collars and satellite tracking of fruit bats in Central Africa. Their results showed that bats could cover 621 miles of territory in a 1-month period. One bat named "Hercules" was recorded to have flown 1180 miles journey in 6 months. The majority of tagged bats were never found, likely due to a high death rate during their travels. Two favored general hypotheses were offered in attempts to explain how a similar Ebola virus strain could appear in two such separate areas,[14,15] since it is highly unlikely that a bat can travel the 2400 mile distance from the Congo to Guinea/Sierra Leone. The first hypothesis is that migrating fruit bats from one area infect other fruit bats in a distant area. Like a relay race, an infected bat in the Congo infects a bat along the way, and this bat continues to pass on the virus to other susceptible bats until one reaches West Africa. A second hypothesis is that some endogenous fruit bats residing in and around the Taï Forest carry the Ebola-Taï Forest strain, whereas other fruit bats in the forest are carriers of the Ebola-Zaire virus. The Taï Forest borders the West coasts of Guinea, Liberia, Sierra Leone, and the Ivory Coast countries and harbors fruit bats, many bird species, pygmy hippopotamus, leopards, and monkeys, including chimpanzees. The relevant fact was that a dead chimpanzee found in the Taï Forest, when autopsied, infected the one and only human known at the time to bear the Ebola virus in West Africa prior to the 2013–15 outbreak. That Ebola virus recovered from the Ebola-Taï Forest had only about a 65% sequence homology with the Ebola-Zaire viruses isolated from humans in the Congo or from those in Sierra Leone or Guinea. Hence, they are very different strains of virus. Perhaps both species of Ebola exist in different bat populations living in the Taï Forest. While many suspect that these bats were a primary source of Ebola, at this time, no full-length sample of the infectious virus has been isolated from these mammals. Rather, fingerprints of antibodies to Ebola and perhaps virus RNA are their current signatures. So the identity of a primary forest host that carries this virus with the capacity to infect humans remains an unsolved mystery.

REFERENCES

1. Kenema Government Hospital. Viral *hemorrhagic fever consortium.* <http://vhfc.org/consortium/partners/kgh> [accessed 20.05.2016].

2. Oskin, Becky. *Awardee Profile—Pardis Sabeti.* Burroughs Wellcome Fund. <http://www.bwfund.org/newsroom/awardee-profiles/awardee-profile-pardis-sabeti>; 2014 [accessed 20.06.16].

3. Pardis Sabeti Computational Geneticist. *National geographic* <http://www.nationalgeographic.com/explorers/bios/pardis-sabeti/>; 2016 [accessed 05.06.16].

4. Cao W, et al. Identification of alpha-dystroglycan as a receptor for lymphocytic choriomeningitis virus and Lassa fever virus. *Science* 1998;**282**:2079–81.

5. Kunz S, et al. Post-translational modification of alpha-dystroglycan, the cellular receptor for arenaviruses by the glycosyltransferase LARGE is critical for virus binding. *J Virol* 2005;**79**:14282–96.

6. ACEGID. *Africa Centre of excellence for genomics of infectious diseases* <http://acegid.org/index.php?active = page>; 2016 [accessed 20.03.16].

7. Mnookin, Seth. *Pardis Sabeti, the Rollerblading rock star scientist of Harvard. Smithsonian*<http://www.smithsonianmag.com/science-nature/pardis-sabeti-the-rollerblading-rock-star-scientist-of-harvard-135532753/>; December 2012 [accessed 02.03.16].

8. *Iranian scientist is one of time's persons of the year* <http://iran-times.com/iranian-scientist-is-one-of-times-persons-of-the-year/>; December 26, 2014 [accessed 20.06.16].

9. Gire SK, et al. Genomic surveillance elucidates Ebola virus origin and transmission during the 2014 outbreak. *Science* 2014;**345**:1369−72.

10. *Ebola (Ebola Virus Disease) transmission.* Centers for Disease Control and Prevention <http://www.cdc.gov/vhf/ebola/transmission/>; July 22, 2015 [accessed 02.03.16].

11. Simon, Scott. *Borders close as Ebola spreads in West Africa* <http://www.npr.org/2014/08/23/342652013/borders-close-as-ebola-spreads-in-west-africa>; August 23, 2014 [accessed 20.06.16].

12. Kolata, Gina. The virus detectives. *The New York Times* <http://www.nytimes.com/2014/12/02/science/factory-direct-virus-analysis.html?_r = 0>; December 1, 2014 [accessed 12.04.16].

13. Holmes EC, Dudas G, Rambaut A, Andersen KG. The evolution of Ebola virus: insights from the 2013−2016 epidemic. *Nat Rev,* 2016;**538**:193−200.

14. Dodds, Kieran. Fruit bats: Africa's greatest mammal migration. *Discover Wildlife* <http://www.discoverwildlife.com/animals/fruit-bats-africas-greatest-mammal-migration>; July 24, 2010 [accessed 04.03.16].

15. Davy, James. Tracking fruit bats may identify regions at greatest risk for ebola epidemic. *Pharmacy Times* <http://www.pharmacytimes.com/news/Tracking-Fruit-Bats-May-Identify-Regions-at-Greatest-Risk-for-Ebola-Epidemic>; September 11, 2014 [accessed 20.06.16].

CHAPTER 8

Ebola's Curse: Impact on the Economics of West Africa

Fear of infection and death overtook West Africans during the Ebola epidemic of 2013–16. This was reflected by Ebola's pivotal role in unraveling the economy of West Africa. As stated by the World Bank, "The largest economic effects of the crisis are not the direct costs (mortality, morbidity, caregiving, and the associated losses of working days), but rather those resulting from changes in behavior-driven by fear which have produced generally lower demands for goods and services and consequently lower domestic income and employment."[1]

An example of that environment of fear was recorded by J. Daniel Kelly in a personal note he published in the journal *Nature* on his experience of going to Africa and witnessing Ebola.[2] "I will never forget the first time I walked into an Ebola isolation ward at Connaught Hospital in Freetown, Sierra Leone. It was 20 August 2014. Inside, eight people thought to have the disease were organized into three patient-care rooms. Patients in the first room appeared to be healthy, and we greeted each other. In the second room, patients barely had the strength to sit. Still, they were able to articulate how they felt. In the last room there were two patients. One was a woman who seemed confused and agitated, and was later confirmed to have the disease. On the other side of the room, a young man was curled into the corner of his bed. He seemed healthy but was terrified. He had been deathly ill when he was admitted three days earlier. He recovered, but had watched Ebola kill two others in that room."

"I could only imagine how I would feel in that situation, watching others get sick and die, wondering if I would be next. Then I considered the deplorable conditions — no visitors were allowed, and a bucket served as a bathroom — and how I, wearing my protective 'spacesuit', must have looked to the curled man. The idea of becoming sick with Ebola in Sierra Leone frightened me…it frightened him too."

Ebola's Curse. DOI: http://dx.doi.org/10.1016/B978-0-12-813888-5.00008-1

"The Sierra Leonean doctor who had supervised the ward had died, and no Sierra Leonean doctor had taken his place. The man was locked in this terrifying environment until someone could draw his blood for testing. "People who think that they might have the disease do not want to spend several days trapped in an isolation unit, away from their families and surrounded by workers in spacesuits. This fear means that patients go to isolation wards only when their symptoms are severe, if they go at all."

The 2013–16 Ebola outbreak in Sierra Leone had and continues to have a drastic negative effect on its economy as well as the economies of Guinea and Liberia. Before Ebola, all these West African countries recorded substantial growth and had overcome most of the economic strain of the recent civil war. In 2013 Sierra Leone and Liberia were ranked second and sixth among the top 10 countries with the highest GDP growth in the world. That GDP subsequently fell as the Ebola curse and scare expanded. World Bank President Jim Yong Kim stated, "Ebola's potential inflicted massive economic costs on Guinea, Liberia and Sierra Leone and on the rest of their neighbors in West Africa."[3]

Of those countries, Ebola cost Sierra Leone the largest decrease in its GDP. Before the Ebola outbreak in 2013, economic growth was at 11.3%, and by end of 2014 it decreased to 8%. Ebola infections spread to 12 of the 13 districts in Sierra Leone with a substantial amount of business disruption and a high rate of deaths.[4] Most involved were the Eastern provinces where Sierra Leone borders Guinea and Liberia. As Sierra Leone's Minister of Agriculture, Joseph Sam Sesay, told the BBC on October 9, 2015, cabinet discussions with President Ernest Bai Koroma revealed that the total economy of the country fell by over 30%.[5] Agriculture, the most important economic section, suffered the greatest impact, since about 66% of the working population were farmers. The police and military set up road blocks, as recommended by the Chief Coordinator for the United Nations Development Program (UNDP) and public health officials, to prevent migrating infected individuals from spreading the Ebola infection to uninfected persons. However, the road blocks also prevented the movement of farmers, other laborers, supplies, shipments of goods, and farm produce.[5]

Farms were simply abandoned, as people who feared contracting Ebola ran away to less disease areas. The result was that no one was left to plant for the next year's and future crops. About 80% of farmers reported that their harvest was smaller than in the previous year.[6] Hardly any food was stored as insurance against future famine. The result was higher food prices and a rise in inflation. Many families could not afford the higher priced food. Heads of around two-thirds of households said they could not purchase enough rice for their needs.[6] The avalanche of difficulties and troubles catapulted. This led to a shortage of money for foreign exchange, a disrupted and angered population, a slowdown in transportation services and halted building operations. Further, in the service sector, the migration out of the area by foreign workers depleted their local spending, which was an important part of the economy in Sierra Leone, Guinea, and Liberia.[7] These cataclysmic economic events led to commercial banks reducing their operations, which included providing loans and capital funding. Bank hours were reduced, usually by half or more, both to save money and because officials feared contact with those infected with Ebola. The tourist industry suffered greatly. Hotels were empty or closed. Staff was laid off. Multiple airlines suspended flights in and out of Sierra Leone and Liberia due to the drop in the tourist trade and dread of viral infection. Limiting air flights compromised deliveries of medical supplies and volunteer health care workers into Ebola-afflicted areas.[8] In all, closure of borders and suspension of air transportation hampered the abilities of Ebola-infected countries to export and import, thereby causing severe economic distresses.

Ebola directly compromised the mining industry in Sierra Leone, Liberia, and Guinea.[9] Mining had accounted for a large part of the country's recent growth potential. Especially hard hit was the world's largest steel maker, ArcelorMittal in Liberia and iron ore mining in Guinea. Closure of mining companies, impairment of services, and weakening of the agriculture sector led to a decline in Guinea's growth rate from the beginning of 2014, predicted to be 4.5% growth but, in 6 months falling to 2.4%.[9]

West African countries rely heavily on agriculture for buttressing their GDP and suffered greatly from that sector's damage. According to the Food and Agriculture Organization, agriculture accounts for 57% of Sierra Leone's GDP, 39% of Liberia's, and 20% of Guinea's.[1] When so many virus-fearing farmers fled their land, Sierra Leone's Agriculture Minister Joseph Sam Sesay stated, "We are definitely expecting a devastating effect not only on labor availability and capacity but we are also talking about farms being abandoned by people running away from the epicenters and going to areas that don't have the disease."[5] With the widespread abandonment of farms, the food supply decreased and food prices increased. David Evans, a senior economist of The World Bank stated, "Three-quarters of the households in Liberia are reporting significant food insecurity"; Liberia's price per rice bag increased from $28 to $35. "Guinea has seen the largest loss in its agriculture industry since it is one of the world's top producers of coca and palm oil."[10]

The epidemic of Ebola virus infection broke out during the season for planting and growing of stable crops such as rice and maize, a severe disruption that lead to pervasive hunger. Chief co-coordinator for the United Nations Development Programme, David McLachlann-Karr elaborated, "We are now coming into the planting season which means a lot of agriculture is not happening, so down the line that will create food shortages and pressures on food prices. We are starting to see a rise in inflation and pressure on the national currency as well as a shortage of foreign exchange."[5] Farmers which comprise the majority of the population suffered greatly. Rural families depend on their crops for their food and income.

Ebola made Sierra Leone, Guinea, and Liberia poorer in other ways. International investors downgraded these countries potential for growth and investments dropped because of the Ebola outbreak. As the World Bank reported on October 7, 2014, less than halfway into the Ebola epidemic, the economic loss due to Ebola virus across Sierra Leone, Guinea, and Liberia would likely reach US$ 32.6 billion by the end of 2015.[11]

The severity of the Ebola epidemic led many non-African countries to become concerned about their exposure to the virus and the virus' transit into their homelands. Heightened control of visitors from West Africa and enhanced national security followed. The international community recognized that the outbreak was a global threat, not just a regional one. To contain the outbreak to Africa and prevent Ebola's spread elsewhere, international funding became available in an effort to control the infection in Sierra Leone, Guinea, and Liberia. The UN stated that US$1 billion was needed to contain the outbreak, even as the World Bank estimated the cost could be billions. Pledges arrived from several countries as well as from multilateral, bilateral, and private organizations. International donors including the World Bank and International Monetary Fund gave approximately US$530 million to the countries hit the hardest. The United States provided $174 million in funding and sent 3000 military troops to build treatment units to assist in containing the outbreak. Money was also used for humanitarian support, fiscal support, screening facilities at borders, airports, and seaports, as well as strengthening the surveillance and treatment capacity of health systems.[12] The African Union, African countries, and the African business community funded efforts to limit Ebola's spread to Nigeria, Cote d'Ivoire, Guinea-Bissau, Senegal, and the Gambia. However, in 2015 a new outbreak of Ebola occurred in Central Africa. This event limited the aid provided by the African Union. Other European and Asian countries as well as Canada pledged funds. Cuba offered more than 460 doctors and health care workers.[13]

Summing up reasons why the Ebola outbreak caused great economic damage, Finance Minister Amara Konneh of Liberia reported that the cause of the fall in GDP was "damage done to mining, agriculture and service industries, loss of foreign workers, border closings, and suspension of international flights."[14] Even further, the effects of spreading Ebola infection unhinged their economy as other governments restricted mobility, trade, and transportation. This was in addition to the substantial direct costs for health care to fight Ebola directly.

Once the Ebola outbreak was recognized, the government of Sierra Leone placed restrictions on people crossing its borders. Cote d'Ivoire and Senegal followed by restricting the movement of people and goods and closing their borders. Countries throughout Africa banned citizens of the three countries most ravaged by Ebola (Guinea, Sierra Leone, and Liberia) from entering their countries and prohibited their citizens from traveling to high-risk areas of Ebola infection. These travel restrictions resulted in many migrant workers losing jobs and businesses along the borders being shut down. Quarantine areas were created where outsiders were not allowed to enter. The United Nations Development Programme (UNDP) urged mandatory placement of road blocks to contain the outbreak. The downside was that road blocks also caused food shortages leading to further hysteria and public protest. Chief co-coordinator of the UNDP stated that "A robust response to quarantining epicenters of the disease is absolutely necessary," but admitted that it has had devastating effects on the agriculture sector.[15]

Cross border trade accounts for from 20% to 75% of the GDP for West African countries.[16] Even so, African countries that traded with Sierra Leone, Guinea, and Liberia reduced or stopped that trade. However, other countries in West Africa where Ebola did not spread such as the Ivory Coast and Senegal also lost trade, exports, and imports due to the fear that their exported products might be infected.

Small local community businesses were negatively impacted because of the restrictions, fear of disease, and decrease in consumers. Businesses that were able to stay open reduced staff and working hours to survive economically, but those reductions decreased business and workers' incomes. The World Bank stated that during the outbreak "half of male heads of households [46%] in Liberia that were working prior to the epidemic remain unemployed, and even more [60%] of female household heads are out of work... and the country's primary productivity has been cut in half."[17]

Tourism was also directly impacted by Ebola's presence. Airlines such as British Airways, Emirates, Air France, Asky Airlines, and Arik Air implemented bans on most flights to and from the affected countries. According to Brookings, "CEOs of 11 firms operating in West Africa have said that some measures, including these travel restrictions, are doing more harm than good and may well be contributing to the humanitarian crisis by blocking crucial trade flows, thereby pushing up the prices of essential foods and medicines." Some examples are the temporary closure of Cameroon's border with Nigeria and the announcement by Kenya Airways that it was suspending all flights to and from Sierra Leone and Liberia.[18]

Ebola's presence severely impacted national and foreign investments. Due to the lack of funds, public spending that was used for the investment in physical capital (machines, equipment) and human capital (labor force) was transferred to other social expenditures. Domestic and foreign investors were cautious and scaled down investing leading to a decline of revenue and stability. According to Voice of America, investor confidence has dropped since the escalation of Ebola infections. Foreign mining corporations such as China Union dramatically scaled down their operations in Liberia.[1] The majority of foreign investments in Sierra Leone, Liberia, and Guinea are directed to their natural resources, many of which are limited or rare to find on other continents. However, when Ebola emerged, many multi-national corporations decided making foreign investments was unprofitable. The increased death rate and rise of quarantined areas caused increased unemployment and made it difficult for corporations to continue running their daily operations. Some miners were afraid to return to work in high-risk districts, quit and did not return to work. Others fled the area. Sierra Leone relies on mining for 17% of its GDP, and Liberia 14% of its GDP.[1] Firms including the Australian mining company, Tawana Resources, and Canadian Overseas Petroleum Limited suspended their operations and sent foreign workers home. This pattern was repeated with other mining companies such as Simandou, Rio Tinto (the world's largest mining company), and London Mining.

The decline in economic activity led to a decline in fiscal revenues as revenues from taxes, tariffs, and customs diminished. The World Bank found that short-term fiscal impacts were large, at $93 million for Liberia (4.7% of GDP); $79 million for Sierra Leone (1.8% of GDP); and $120 million for Guinea (1.8% of GDP).

The WHO and international community, once alerted, attempted to help the distressed countries overwhelmed by the economic and human devastation Ebola viruses wreaked, but their commitment was overall insufficient and tardy. Thus, poor public health knowledge, delay in government and international response, and limited cooperation were all complicit in allowing a disease whose spread was preventable to go from a few hundred to 28,000 cases, leading to severe economic difficulties for the Ebola-infected countries.

REFERENCES

1. Sy, Amadou, and Amy Copley. *Understanding the economic effects of the 2014 Ebola outbreak in West Africa*. The Brookings Institution <http://www.brookings.edu/blogs/africa-in-focus/posts/2014/10/01-ebola-outbreak-west-africa-sy-copley>; October 1, 2014 [accessed 20.06.16].

2. Kelly J Daniel. Making diagnostic centers a priority for Ebola crisis. *Nature* 2014;**513**:145.

3. Geewax, Marilyn. *World bank says Ebola could inflict enormous economic losses*. NPR <http://www.npr.org/sections/thetwo-way/2014/10/08/354599549/world-bank-says-ebola-could-inflict-enormous-economic-losses>; October 8, 2014 [accessed 12.03.16].

4. Jallanzo, Ahmed. *Ebola: economic impact could be devastating*. World Bank<http://www.worldbank.org/en/region/afr/publication/ebola-economic-analysis-ebola-long-term-economic-impact-could-be-devastating>; August 2014 [accessed 22.04.16].

5. Hamilton, Richard. Ebola crisis: the economic impact. *BBC News* <http://www.bbc.com/news/business-28865434>; August 21, 2014 [accessed 13.06.16].

6. Ebola hurts more than the sick: world bank. *NBC News* <http://www.nbcnews.com/storyline/ebola-virus-outbreak/ebola-hurts-more-sick-world-bank-n284421>; January 12, 2015 [accessed 20.06.2016].

7. Odutayo, Armide. The Ebola virus disease: problems, consequences, causes, and recommendations. *E-International Relations* <http://www.e-ir.info/2015/04/22/the-ebola-virus-disease-problems-consequences-causes-and-recommendations/>; April 22, 2015 [accessed 20.06.16].

8. Kenya airways to suspend flights to freetown, Monrovia due to Ebola. *Reuters* <http://www.reuters.com/article/us-health-ebola-kenya-airways-idUSKBN0GG0F520140816>; August 16, 2014 [accessed 13.05.16].

9. Neate, Rupert. Mining company at centre of fight against Ebola in Sierra Leone goes bust. *The Guardian* <http://www.theguardian.com/world/2014/oct/16/london-mining-fight-ebola-sierra-leone-goes-bust>; October 16, 2014 [accessed 13.06.16].

10. *The socio-economic impacts of Ebola in Liberia.* World Bank <http://www.worldbank.org/content/dam/Worldbank/document/Povertydocuments/socio-economic-impact-of-ebola-on-households-in-Liberia(final).pdf>; November 19, 2014 [accessed 12.05.16].

11. *The economic impact of the 2014 Ebola epidemic: short and medium term estimates for West Africa.* World Bank <http://www.worldbank.org/en/region/afr/publication/the-economic-impact-of-the-2014-ebola-epidemic-short-and-medium-term-estimates-for-west-africa>; October 8, 2014 [accessed 13.06.16].

12. UN: nearly $1 billion needed to combat Ebola outbreak. *UN News Center* <http://www.un.org/apps/news/story.asp?NewsID = 48728#.VxmRU7_Vuho>;. September 16, 2014 [accessed 13.06.16].

13. Freeman, Colin. Cuban doctors take leading role in fighting Ebola." *The Telegraph* <http://www.telegraph.co.uk/news/worldnews/ebola/11375422/Cuban-doctors-take-leading-role-in-fighting-Ebola.html>; January 29, 2015 [accessed 13.06.2016].

14. E.W. Ebola's economic impact. *The Economist* <http://www.economist.com/blogs/baobab/2014/09/costs-pandemic>; September 3, 2014 [accessed 13.06.16].

15. Gallagher, James. Ebola response lethally inadequate, says MSF. *World in Struggle* <http://worldinstruggle.blogspot.com/2014/09/west-africa-ebola-plague-caused-by-imf.html>; September 2, 2014 [accessed 13.05.16].

16. *Harmonizing policies to transform the trading environment. Assessing regional integration in Africa VI*; October 3, 2013 [accessed 22.05.16]. http://dx.doi.org/10.18356/5d7bf72c-en.

17. *Ebola hampering household economies across Liberia and Sierra Leone.* World Bank <http://www.worldbank.org/en/news/press-release/2015/01/12/ebola-hampering-household-economies-liberia-sierra-leone>; January 12, 2015 [accessed 13.05.16].

18. Mjamba, Khanyo Olwethu. *Cameroon closes border with Nigeria over Ebola.* This Is Africa <http://thisisafrica.me/cameroon-closes-border-nigeria-ebola/>; August 18, 2016 [accessed 13.06.16].

CHAPTER 9

Ebola's Scorecard: Failure of the WHO and the International Community

The international health community and its institutions made a slate of errors, each of which prolonged, helped to spread, and continued the Ebola epidemic from 2013 to its current status in 2016. Their failure took the form of responding too slowly, too inefficiently, and too ineffectively. When they did respond international organizations such as the World Health Organization (WHO), World Bank, and United Nations (UN) failed in their communications among one another, a major cause for delay of effective action needed to contain the viral disease. Nongovernmental organizations (NGOs) clashed with the WHO and World Bank; the result was confusion about what was going on, what to do, or who was to do what. Multinational corporations initially contributed little of substance. Here, we describe the faults that caused such damage for the purpose of recommending steps to correct the errors and ensure that a catastrophe like this does not occur again.

The WHO has its share of blame and rightly so. Its chartered role is to protect the health of our world's human populations, and its task is to live up to that purpose. It's major malfunction in the Ebola epidemic was its failure to understand the public health disaster as it was unfolding and thus not acting early enough to limit the spread of Ebola virus infection. During the initial months of Ebola's debut, the WHO was slow in acknowledging that a unique infectious disease was advancing and killing people. Throughout this indolent period, Ebola continued to spread from Guinea across neighboring countries, primarily Sierra Leone and Liberia. With optimal public health control, the numbers of infections and deaths might have been 10-fold fewer or less, i.e., 28,000 to 2800 and 11,300 to 1300. Compared to the WHO's successes in handling prior epidemics of H1N1 and H5N1 influenza and Severe Acute Respiratory Syndrome (SARS), the Ebola epidemic was largely ignored and handled poorly. The major cause of this failure by the WHO was likely its administration at that time. Neither

Ebola's Curse. DOI: http://dx.doi.org/10.1016/B978-0-12-813888-5.00009-3

their personnel nor that of the international disease control community was sufficiently decisive in understanding, acting, or supervising the outbreak. "We can mount a highly effective response to small and medium-size outbreaks, but when faced with an emergency of this scale, our current systems — national and international — simply have not coped," stated WHO Director-General Margaret Chan, Deputy Director-General Anarfi Asamoa-Baah, and the organization's regional directors in a joint statement on April 16, 2016.[1] But, of course, the earlier an infection is contained, the less likely it will spread from a small to a large problem. The WHO admitted it was ill pre-pared. "We have taken serious note of the criticisms that the initial WHO response was slow and insufficient, we were not aggressive in alerting the world ... we did not work effectively in coordination with other partners, there were shortcomings in risk communications and there was confusion of roles and responsibilities".[2] In contrast, the WHO led by a different administration in 2002—04 acted forcefully and correctly when faced with the outbreak and possible pandemic effect of SARS.[3]

The WHO was also defective in monitoring the Ebola outbreak. A critique by a group of 20 experts from the Harvard Global Health Institute and the London School of Hygiene and Tropical Medicine found that "The lack of capacity in Guinea to detect the virus for several months was a key failure, allowing Ebola eventually to spread to bordering Liberia and Sierra Leone, underscoring inadequate com-munication and arrangements between governments and the WHO to share, validate, and respond robustly to information on outbreak."[4] Indeed, after Ebola was initially identified, it still spread through the capital cities of Guinea and Liberia, and within 2 months appeared in other major cities and their international airports. Without protocols in place for identification of Ebola, the virus rapidly spread. The *RT International* report of November 23, 2015 stated, "Without any approved drugs, vaccines or rapid diagnostic tests, health workers struggled to diagnose patients and provide effective care. Without suf-ficient protective gear, and initially without widespread understanding of the virus, hundreds of health workers themselves became ill and died."[4]

In summary, early in the course of the Ebola infection, before its massive outbreak, Doctors Without Borders warned the WHO about the potential threat. This evaluation, despite its highly qualified source, was originally disputed by the WHO. As a result, actions to fight the infection and arrange for humanitarian aid were delayed. Not until August 2014, a good 8 months after the initial Ebola cases emerged in Guinea, did the WHO begin to take action. The Harvard Global Health Institute's report called for greater accountability and transparency within all global health institutions and remarked that the WHO should respond to freedom of information requests.

In the West African countries, consistently poor health care and lack of adequate infrastructure were major factors in the increasing difficulty of addressing public health concerns and medical emergencies. By comparison, in Boston, at Peter Bent Brigham Hospital—part of Harvard's medical complex, more physicians worked on the second floor alone than in all of Liberia. Most of the healthcare staff in countries overrun with Ebola virus infections were not sufficiently trained to respond. They often lacked even the basic materials required for treatment and had insufficient knowledge or equipment to protect themselves from contamination. The exceptions in some areas included Kenema Government Hospital (KGH) in Sierra Leone and centers where Doctors Without Borders were located. For example, Dr. Mariano Lugli, a deputy director of operations for Doctors Without Borders, did respond to an early incidence of Ebola virus infection. Working in remote forests of Guinea during March, 2014, when the outbreak spread to Guinea's capital, Conakry, Lugli set up a healthcare receiving and treatment clinic. Although Lugli was met by a foreign medic and logistician sent by the UN health agency, he never saw or met a WHO official who was responsible for handling this escalation of the outbreak. Lugli elaborates, "In all the meetings I attended, even in Conakry, I never saw a representative of the WHO. The coordination role the WHO should be playing, we just didn't see it. I didn't see it the first three weeks and we didn't see it afterwards."[5]

Because so many patients and their healthcare providers had already died and those not yet infected feared the same fate, many hospitals were shut down and abandoned. Hundreds of patients remained waiting in front of nonfunctional hospitals in the hope of being admitted and treated. The WHO received extensive criticism for taking too long to provide and organize the flow of physicians, healthcare workers, protective clothing, and even fluids. Without leadership by the WHO in pursuing outside governments and philanthropies to establish isolation centers, surveillance, and laboratory capacity in West Africa, the local governments turned school classrooms into holding centers for those suspected of carrying the viral disease. Even so, neither bedding nor basic medical equipment were available. For the most part this effort turned out to be useless.

In similar straits, Dr. Melvin Korkor, in charge of Phebe Hospital in Liberia, spoke of repeated delays in receiving much needed materials, none of which was available in the region. Many patients did not receive basic medications. Supplies of test tubes, gowns, and fluids were depleted. Medical staff lacked basic safety equipment and sterile latex gloves, without which their hands were unprotected while treating patients exposing these frontline health providers to the virus.[6] The end result was a high mortality rate among the care givers. Dr. Korkor considered himself "reborn" after surviving Ebola infection. The lack of doctors and trained medical workers in West Africa played a role in the spread of Ebola. Liberia has only one doctor for every 100,000 people, whereas Sierra Leone has two. In comparison, the United States has 245 doctors per 100,000 individuals. As hundreds of local doctors died in African communities, the Ebola outbreak escalated. Yet the WHO knew about the lack of health infrastructure in these countries, and one of their priorities should have been to plan support and enhancement of the healthcare network. Their responses should have been more vigorous.[7]

Failing to notify the global community about the rapid spread and danger of the Ebola outbreak was a major error. Ashish Jhna, director of the Harvard Global Health Institute, stated "People at WHO were aware that there was an outbreak that was getting out of control by Spring, and yet it took until August to declare a public health emergency."[8] The Harvard Institute also accused the WHO of enabling "immense human suffering, fear, and chaos" as a result of their delayed response to the epidemic. A vivid example of poor management was the handling of early blood samples taken from infected patients to determine if they had Ebola. Some samples were shipped to laboratories where they were not examined immediately. Others sent to Paris, France, could not be tested at the recipient institution due to technical difficulties and had to be re-routed to Lyon, 250 miles away.

Thus, analysis of whether an individual was infected and should be quarantined was delayed, another administrative failure.

What was the reason for the problems encountered by WHO and for its delayed actions? According to the WHO, one major obstacle was their concerns about political opposition from West African leaders. Many were cautious about taking aid because they mistrusted the source, a reflection of past exploitation by the West. The WHO, instead of working vigorously to resolve this difficulty, providing education, and reinforcing communication, became politically correct. When they should have announced that a major infectious outbreak from a deadly virus would hurt these countries' economies, the WHO did nothing to improve those relationships. Sensitive cultural differences made the WHO leery of disrupting any country's governance without a consensus. Unfortunately, regional culture most often trumps science and reason. The WHO bowed primarily to political pressure rather than health concerns. Critics have said, and we concur, that the WHO should have understood that traditional and natural practices in the region stood in the way of effective mechanisms to contain the virus, and the WHO had a moral obligation to act as educators, to organize teachers, and to share scientific knowledge of what the outcome would be not only to political leaders of government but most importantly to local tribal leaders.

Inadequate funds and, when available at all, poorly used were part of the problem. The leader of Doctors Without Borders, Ebola response team, Christopher Stokes, said it was "ridiculous" that volunteers working for his charitable group were bearing the brunt of care in the most severely affected countries and that international efforts will not have any effect for more than a month.[9] As a defense for not arranging to provide sufficient funding to control Ebola' destruction, Director-General Margaret Chan of the WHO explained that the WHO is not an implementation agency for outbreak response:

"First and foremost, people need to understand WHO. WHO is the UN specialized agency in health. And we are not the first responder. You know, the government has first priority to take care of their people and provide health care. WHO is a technical agency. So this is how we provide services. We are not like international National Government Organizations (NGOs), for example Doctors Without Borders, Red Cross, Red Crescent or local NGOs who are working on the ground to provide, you know, direct services".[10]

However, the WHO has the pulpit on the world stage to organize international efforts. In this it was negligent. WHO should have been crisp in effective decisions to fight Ebola. It was not. It was wishy-washy, as its consensus-building approach and political correctness show. Instead, a bureaucracy that wishes to be admired and retained did not stir the pot by being decisive. We believe that the WHO should act aggressively once the science is known during potential major health disasters. They should be on the frontline to mobilize world support for control of such infectious diseases. Previously, they followed that kind of aggressive policy to combat the first 21st century pandemic, SARS, but that was at a different time and under different leadership.[3] A safe political bureaucracy should not be the game plan for setting up and funding the WHO.

Additional problems had hindered any effective response to Ebola. Bureaucracies of many countries blocked or delayed responses to the outbreaks by denying visas to scientists, doctors, and healthcare workers who tried to cross their borders to help the victims.

The World Bank, an international agency that provides loans to developing countries, also made errors during the Ebola outbreak. Predominant was its delayed response and poor cooperation with the WHO. The World Bank had opportunities early during the outbreak to meet the requests of scientists and government officials for finances to cover basic, necessary operations but chose to respond later. Oxfam, an international confederation of 18 NGOs working with partners in over 90 countries, criticized the World Bank for its failure to invest enough in the region's healthcare infrastructure. Jim Kim, president of the World Bank, admitted his institution's failures, "We should have done so many things. Healthcare systems should have been built. There should have been monitoring when the first cases were reported. There should have been an organized response."[11]

Many critics have pointed out that the delay in response to spreading Ebola virus infection was caused, in large part, by the lack of cooperation and disagreement between the WHO and World Bank on a plan of action. Kim admitted that failure and stated, "The most important thing is to stop arguing about what is or is not possible and to get on with doing what is needed."[12]

The African Development Bank played a large part in contributing funds to the beleaguered communities. This contribution was supported by other NGOs, international organizations, and countries. In 2014, the African Development Bank provided a total of 223 million US dollars to Guinea, Liberia, and Sierra Leone. That bank additionally collaborated with the WHO to supply additional resources including medicines, equipment, and emergency training. In addition to the promised and paid contributions, the African Development Bank also established two new post-crisis operations to lessen and prevent the instability and chaos caused by the Ebola outbreak. These operations included the establishment of an African Centre for Disease Control, and a post-Ebola Livelihoods Restoration Project.[13]

An important goal of the UN is to provide humanitarian aid in times of famine and natural disasters. However, this time, the UN did not live up to its responsibilities. According to Doctors Without Borders, the UN had minimal impact on the epidemic regardless of their international pledges and deployments of staff. However, David Nabarro, a medical doctor who organized and led the UN mission to alleviate Ebola, disagreed "I am absolutely certain that when we look at the history, this effort that has been put in place will have been shown to have had an impact, though I will accept that we probably won't see a reduction in the outbreak curve until the end of the year."[14] All told, Doctors Without Borders was the frontline team in the fight against Ebola despite their frustration with the lack of support in terms of action and supplies they needed during the epidemic. That organization ran the majority of Ebola treatment facilities across the region, providing over 700 of the 1000 beds available. The UN was not the source of frontline defenders fighting the epidemic.

The international community of nations was also deficient in the humanitarian effort it should have supplied to West Africa. Many countries failed to commit needed medical or financial resources toward alleviating the outbreak. Large and wealthy countries such as China, Russia, and Saudi Arabia barely contributed anything. In contrast, a smaller country such as Cuba and a rich country such as Sweden participated in larger, prominent, and more effective ways. The majority of donations provided came from the United States, United Kingdom, and Germany. The United States funded over a third of the UN relief fund. Several banks and philanthropists were also active contributors.

The United States and United Kingdom provided the most support. Shamefully, responses from the rest of Europe and the European Union were limited and quite unsatisfactory. For example, France pledged $89.7 million; $44.85 million in direct bilateral aid and $44.85 million to multilateral institutions, a meager contribution from its $2.61 trillion economy. Northern European countries donated a significant amount more than Germany, which has one of the largest economies in Europe. Germany agreed to donate $13.37 million, contributed to international aircraft and to building a field house with 300 beds in Guinea. The Netherlands contributed the most among its regional neighbors.[15] Canada generously pledged over $100 million and sent supplies.

Although possessing a small economy, Cuba played a significant role in providing medical staff. Cuba sent substantial human resources in that more than 460 doctors and medical staff went to ease the crisis in West Africa. Other South American countries also donated: Brazil pledged $450,000 to the WHO along with donating five supply kits, each of which can protect 500 workers from Ebola. Chile and Columbia have also donated funds.[16]

To better understand contributions and commitments of countries, it is critical to examine how much countries donated relative to their economy. That is donations by GDP.

Although the United States and United Kingdom have pledged the most funds among all countries, some smaller countries surprisingly pledged more money relative to their GDP. Along with the United Kingdom and United States, Canada as well as Australia and Japan stand out as having contributed the most relative to their economy. Other countries such as China, Russia, Italy, France, and Germany were poor donors, and why they were not more involved is unclear. The World Food Program (WFP), in particular, lashed out at Beijing's wealthy. "Where are the Chinese billionaires and their potential impact? Because this is the time that they could really have such a huge impact," said Brett Rierson, WFP representative in China.[17] Private donors as well as the government of China afforded only meager responses, considering that China is a major investor in Africa. Although it was in that nation's political interest to expand their influence and potential contacts among the African countries, the government and private sector in China contributed a total of only

$8.3 million to the UN main Ebola relief fund (compared to more than $200 million from the United States). Of that $8.3 million, only $4.89 million came from the Chinese government. However, China did expand its medical staff in Sierra Leone to 50 laboratory members and promised to contribute another $34 million, but as of this writing has yet to fulfill that obligation. Beijing announced that it would donate up to $4 million to the WHO.[18]

In addition to governments, some NGOs, multinational corporations, and other international organizations played a smaller than expected role during the Ebola crisis. However, other NGOs, specifically philanthropists and wealthy individual charities, made considerable funding available. Over 60 NGOs, foundations, and charities have provided much needed funding for equipment and supplies.[19] Among the major contributors were The Bill and Melinda Gates Foundation, Oxfam, Save the Children, Paul G Allen Family Foundation, Silicon Valley Community Foundation, and the Ikea Foundation.

Multinational corporations, also known as Corporate Enterprises, were slow to respond. During past natural disasters, multinational corporations responded far more quickly and generously. Many companies that rely on natural resources in West Africa offered little to no help. For example, the cocoa industry relies heavily on West Africa products. Seventy percent of the world's supply comes from this region. Large multinational corporations like Nestle, Mars Chocolate, and Hershey's have donated a meager $700,000 to their Cocoa Foundation to support the effort against Ebola.[20]

In past natural disasters and global emergencies, the international community also responded far more magnanimously than recorded for the 2013–16 Ebola epidemic. Similarly, international organizations such as the WHO, World Bank, and UN appeared to do too little. At the end of the day, it seems that everyone insisted something should be done, but few took action. According to the internal WHO report, "Nearly everyone involved in the outbreak response failed to see some fairly plain writing on the wall. A perfect storm was brewing, ready to burst open in full force."[21]

If all this went wrong, then how can the world community plan so this disaster does not happen again? Put another way, "Those who fail to learn from history, are doomed to repeat it." Further, as populations in West Africa increase and more humans breach the forest area, in all likelihood a new Ebola epidemic will occur. So a storm is brewing; if so, what recommendations are needed for preparations, devising a system for global warning, and implementing a response system? The stated purpose of the foregoing review is to contemplate the management of global health crises, and our suggestions follow.

First, better cooperation and communication among international agencies, governments, and NGOs are required. To accomplish this task, the development, commitment, and use of a single global institution with the responsibility for natural and environmental epidemics may be essential. Would it be more effective to create a new institution, rather than giving the authority to an existing global institution such as the CDC, WHO, or UN? Certainly these and other currently operational institutions can and have played important roles in the past. The best argument for a new institution is to generate one with a single focus. The current global institutions have multiple responsibilities and priorities. Creating a solely "nonpolitical" and lean institution whose primary goal is for the prevention and recovery of global and natural health disasters would eliminate the lengthy and bloated bureaucratic process so that action would be taken quickly and efficiently. One new institution currently being put in place to meet these responsibilities and could be up to the task is the Global Virus Network (GVN). GVN represent centers of excellence in medical virology. It's work is to understand, prevent, and eradicate viral disease threats to mankind. GVN or a similar type of focused institution might also have full authority over the distribution of raised assets and a reserve fund.

The last category brings us to the second recommendation, setting up an emergency fund. Acquiring a fund that already has pledged donations and resources would allow immediate action. A rapid action plan would impact future outbreaks of disease by quickly down-modulating their spread. This fund would be used for emergency preparation, recruitment, shipping, doctors and healthcare workers, as well as training populations in third-world areas where outbreaks are likely to occur. Also, necessary medicines and medical equipment should be stocked, stored safely, and available. Funds for research and facilities should be pre-arranged. One would have to gather international support and trust in creating a new global institution for this goal.[22] The need is to convince the world's sources of wealth to fund an enterprise that manages epidemic-specific activities and to allocate part of their GDP to this effort. The world is getting smaller. Diseases like Ebola, Zika, yellow fever, and Lassa are more easily transported by air flight than ever before. Thus, the threats of these epidemics are today's reality not only in countries where outbreaks occur, but in all countries engaged in global trade.[23]

The third recommendation is the need to expand global investment toward international health and natural disasters. Sadly, only a few countries have met their commitments under the International Health Regulations created by the UN after the SARS outbreak. Realistically, the interests of many countries and institutions are best served by allocating assets toward elimination/control of epidemics despite the fact that they usually occur far from Western countries. The global community must recognize that these "exotic" African diseases readily arrive to distant shores of non-African countries carried by travelers incubating the infectious agent. This is not theoretical but actually occurred in the United States, United Kingdom, and elsewhere with the transport of Ebola, Lassa, and currently Zika viruses. Further, there is the possibility of spread of viruses like Ebola and Lassa by bioterrorism. In the past (1918–19) an influenza epidemic infected over 5% of the world's population and killed approximately 2% (over 50 million people).[24] The World Bank projected that the cost of inaction of a worldwide influenza epidemic would reduce global wealth by over $3 trillion.[25]

A fourth issue is represented by the failure of Ebola containment. This was, in large part, due to the lack of a universal and robust disease surveillance system. Ideally, such surveillance systems would be part of a global public health network. The ability to perform in-depth, rapid sequencing to identify the virus in question in hours or at least a day is now available. This sequencing of individual blood samples during ongoing epidemics is now possible even in remote areas, where the majority of outbreaks occur.[26] Surveillance helps increase effective communication among global institutions, countries, and citizens and would greatly decrease the impact of any epidemic.

In the past, self-serving politics and flawed policies were the cause of delays that killed thousands; therefore, the fifth issue of reforms in global health care is to avoid bureaucratic issues that prevent the speedy and worldwide release of data that forecast epidemics. The essential mandate is that, once data are obtained and verified, they are released immediately, not withheld for personal gain or credit or by countries wishing to mislead travelers and businesses.

The fifth recommendation in this world health plan is that essential investments should be made to train teams of doctors, medical staff, healthcare workers, and researchers. During the 2013−16 Ebola outbreak there was an enormous demand for doctors, nurses, and medical staff that was never met. Training in global health should be a component of infectious disease training not only in Ebola-susceptible countries but also in the medical schools and residencies of European, American, Canadian, South American, and Asian countries. The West African countries where Ebola infection prevailed still lack doctors. The dearth of basic equipment in Ebola-afflicted West Africa contributed to the large number of deaths and unsafe medical practices among healthcare workers. In addition, protective equipment—uniforms, gloves, and headgear—should always be available. Having trained individuals would speed up the control process and limit the spread of the infecting agent. During the 2013−16 Ebola outbreak, medical volunteers did not step forward in force to help until 2−3 months after the Ebola outbreak. An enhanced medical team could have prevented escalation of the epidemic. The Bill and Melinda Gates Foundation suggests, "We need to invest in better disease-surveillance and laboratory-testing capacity, for normal situations and for epidemics. Routine surveillance systems should be designed in such a way that they can detect early signs of an outbreak beyond their sentinel sites and be quickly scaled up during epidemics...and the data derived from such testing need to be made public immediately. Many laboratories in developing countries have been financed by the polio-eradication campaign, so we will have to determine what capacities will be needed once that campaign is over."[23]

Finally, an improved education plan is an absolute requirement. Public understanding of how a virus spreads, the value of quarantine, and basic public health measures could stop the spread of an ongoing epidemic. An important lesson taught by the 2013−16 Ebola outbreak is that the susceptible populations must be educated in what and why public health measures are needed. The local heads of countries, districts, and most importantly village leaders, must understand how the disease travels, so that they can lead and guide their population in turn. Surveillance in all areas including rural sites should be unrelenting along with acceptance of quarantines, appropriate treatments, and safe burial practices. Along with these recommendations, courage, grit, and prayer should provide the format and strength to complete the goals of successful medical control of future epidemics.

Lastly, what have we learned from the Ebola outbreak in West Africa 2013−16? The science and the underlying advances, management successes, along with the cultural and bureaucratic difficulties and failures come clear. But more than that, the numerous tales of humanity, the human story of native and foreign influx of healthcare workers, the role of KGH staff, Drs. Khan, Sabeti, and Garry, Doctors Without Borders, CDC, missionary hospitals as well as volunteers to fight the epidemic and care for the ill stand out front. Their stories reaffirm that all people are connected to other people and dependent on other people. These events in West Africa in 2013−16 resonate with lines written over 390 years ago by the English poet John Donne in his *Devotions upon Emergent Occasions*

No man is an island,
Entire of itself.
Each is a piece of the continent,
A part of the main.
If a clod be washed away by the sea,
Europe is the less.
As well as if a promontory were.
As well as if a manor of thine own
Or of thine friend's were.
Each man's death diminishes me,
For I am involved in mankind.
Therefore, send not to know
For whom the bell tolls,
It tolls for thee.

John Donne

REFERENCES

1. McSpadden, Kevin. WHO acknowledges failings of Ebola response. *Time*; 20 April 2015 [Web. 21.06.16].

2. World Health Organization admits it failed in handling Ebola. *NY Daily News* <http://www.nydailynews.com/life-style/health/world-health-organization-admits-failed-handling-ebola-article-1.2191334>; April 20, 2015 [accessed 21.06.16].

3. Oldstone MBA. *Severe acute respiratory syndrome (SARS): the first pandemic of the 21st century. Viruses, plagues, & history*. NY: Oxford University Press; 2010.

4. WHO 'failed to Alert' Global Community about Ebola Outbreak Allowing Virus to Spread Further — Panel. *RT International* <https://www.rt.com/news/323113-ebola-outbreak-who-failure/>; November 23, 2015 [accessed 21.06.16].

5. Goldberg, Eleanor. Ebola aid workers shocked by WHO's 'amateurism' in response to outbreak. *The Huffington Post* <http://www.huffingtonpost.com/2014/10/06/who-poor-response-ebola_n_5933866.html>; October 5, 2014 [accessed 21.06.16].

6. Hinshaw, Drew. Ebola virus: for want of gloves, doctors die. *WSJ* <http://www.wsj.com/articles/ebola-doctors-with-no-rubber-gloves-1408142137>; August 16, 2014 [accessed 21.06.16].

7. Duff, Michael. Ebola takes big toll on already poor health care. <http://www.redding.com/news/ebola-takes-big-toll-on-already-poor-health-care-ep-585823658-362256141.html>; August 30, 2014 [accessed 21.06.16].

8. Colen BD. An indictment of Ebola response. *Harvard Gazette* <http://news.harvard.edu/gazette/story/2015/11/an-indictment-of-ebola-response/>; November 22, 2015 [accessed 21.06.2016].

9. Ebola crisis: no impact from pledges of help, MSF Says. *BBC News* <http://www.bbc.com/news/world-africa-29656417>; October 17, 2014 [accessed 21.06.16].

10. Fink, Sheri. W.H.O. leader describes the agency's Ebola operations. *The New York Times* <http://www.nytimes.com/2014/09/04/world/africa/who-leader-describes-the-agencys-ebola-operations.html>; September 4, 2014 [accessed 21.06.16].

11. Elliott, Larry. Ebola crisis: global response has 'failed miserably', says World Bank Chief. *The Guardian* <http://www.theguardian.com/world/2014/oct/08/ebola-crisis-world-bank-president-jim-kim-failure>; October 9, 2014 [accessed 21.06.16].

12. Global 'failure' to grip Ebola crisis criticised by World Bank President | World News |. *NNWnet National News Wire* <http://nnw.net/global-failure-to-grip-ebola-crisis-criticised-by-world-bank-president-world-news/>; October 08, 2014 [accessed 21.06.16].

13. *Ebola*. African Development Bank Group <http://www.afdb.org/en/topics-and-sectors/topics/ebola/>; 2006 [accessed 21.06.16].

14. Ebola crisis: UN Envoy rejects criticism of agency's response. *BBC News* <http://www.bbc.com/news/world-africa-29672179>; October 18, 2014 [accessed 21.06.16].

15. *Action taken by the Netherlands and the International Community to tackle Ebola.* Information from the Government of The Netherlands <https://www.government.nl/topics/ebola/contents/action-taken-by-the-netherlands-and-the-international-community-to-tackle-ebola> [accessed 21.06.16].

16. Szklarz, Eduardo. Latin American health officials prepare to fight Ebola. *Diálogo* <http://dialogo-americas.com/en_GB/articles/rmisa/features/2014/10/16/feature-01?source = most_viewed>; October 16, 2014 [accessed 21.06.16].

17. Rajagopalan, Megha. China's companies, billionaires must step up to fight Ebola: WFP. *Business Insider* <http://www.businessinsider.com/r-chinas-companies-billionaires-must-step-up-to-fight-ebola-wfp-2014-10>; October 20, 2014 [accessed 21.06.16].

18. Sanchez, Raf. What countries have pledged to fight Ebola... and how much they've paid into the fund. *The Telegraph* <http://www.telegraph.co.uk/news/worldnews/ebola/11179135/What-countries-have-pledged-to-fight-Ebola...-and-how-much-theyve-paid-into-the-fund.html>; October 22, 2014 [accessed 21.06.16].

19. *Non-Governmental Organizations responding to Ebola.* USAID Center for International Disaster Information CIDI <http://www.cidi.org/ebola-ngos/#.V0yr_4-cGrM>; 2015 [accessed 21.06.16].

20. Rooney, Ben. UN asks $1 billion for Ebola, gets $14 million so far. *CNNMoney* <http://money.cnn.com/2014/10/17/news/un-ebola-funding/index.html>; October 21, 2014 [accessed 21.06.16].

21. Cheng, Maria. Ebola outbreak: WHO admits it botched early attempt to stop disease. *CBC News* <http://www.cbc.ca/news/world/ebola-outbreak-who-admits-it-botched-early-attempt-to-stop-disease-1.2802432>; October 17, 2014 [accessed 21.06.16].

22. Rull, Monica, Ilona Kickbusch, and Helen Lauer. *International Development Policy Revue Internationale De Politique De Développement*. Policy Debate <https://poldev.revues.org/2178#tocto2n3>; June 2, 2015 [accessed 21.06.16].

23. Gates Bill. The Next Epidemic Lessons from Ebola—NEJM. *New Engl J Med* 2015. [accessed 21.06.16] <http://www.nejm.org/doi/full/10.1056/NEJMp1502918>.

24. Oldstone MBA. *Viruses, plagues, & history*. New York, NY: Oxford University Press; 2010.

25. The World Bank. *Pandemic risk and one health* <http://www.worldbank.org/en/topic/health/brief/pandemic-risk-one-health>; October 23, 2013.

26. Holmer Edward, Dudas Gytis, Rambaut Andrew, Andersen Kristian. The evolution of Ebola virus: Insights from the 2013–2016 epidemic. *Nat Rev* 2016;**538**:193–200.

FURTHER READING

Sanchez, Raf. What countries have pledged to fight Ebola... and how much they've paid into the fund. *The Telegraph* <http://www.telegraph.co.uk/news/worldnews/ebola/11179135/What-countries-have-pledged-to-fight-Ebola...-and-how-much-theyve-paid-into-the-fund.html>; October 22, 2014 [accessed 21.06.16].

Since our book was written, results for an Ebola vaccine developed by Merck and New York Genetics Corporation tested in a 4160 patient trial in Guinea became available with much euphoria. The vaccine study was somewhat unique from other field studies as instead of randomizing susceptible individuals, medical personnel identified outbreak areas and vaccinated people in rings around the original cases. The first ring was around the Ebola cases, and subsequent rings or clusters farther away were vaccinated 21 days later and so on. The vaccine was genetically engineered. The Ebola glycoprotein (protein responsible for attaching the virus to the cell it infects) replaced the glycoprotein of vesicular stomatitis virus (VSV). The vaccine was a recombinant VSV expressing only the glycoprotein gene of Ebola and allowed replication of the Ebola glycoprotein. This would lead to induction of neutralizing antibodies, potentially cytotoxic T cells only against the Ebola glycoprotein but no responses to the other seven proteins of Ebola. Unfortunately these immunologic parameters were not obtained in the trial and the results of the trial have been controversial (https://www.nap.edu/catalog/24739/integrating-clinical-research-into-epidemic-response-the-Ebola-experience; Henao-Restrepo et al. *Lancet* 2017;**389**:505−18). Developers and administrators of the vaccine claimed close to 100% effectiveness from contracting Ebola once vaccine protection kicked in. In contrast, a panel of the US National Academy of Medicine came to a different conclusion. They found due to varying ways the data was analyzed and trial conducted effectiveness of the vaccine was unclear although the vaccine did prevent infection in some. Thus the final verdict on whether a successful Ebola vaccine is on hand is unclear.

Note: Page numbers followed by "*f*" and "*t*" refer to figures and tables, respectively.

Printed in the United States
By Bookmasters